17

HINDSIGHT AND
POPULAR ASTRONOMY

HINDSIGHT AND POPULAR ASTRONOMY

Alan B Whiting
University of Birmingham, UK

World Scientific

NEW JERSEY · LONDON · SINGAPORE · BEIJING · SHANGHAI · HONG KONG · TAIPEI · CHENNAI

Published by

World Scientific Publishing Co. Pte. Ltd.

5 Toh Tuck Link, Singapore 596224

USA office: 27 Warren Street, Suite 401-402, Hackensack, NJ 07601

UK office: 57 Shelton Street, Covent Garden, London WC2H 9HE

British Library Cataloguing-in-Publication Data
A catalogue record for this book is available from the British Library.

HINDSIGHT AND POPULAR ASTRONOMY

ISBN-13 978-981-4307-91-8
ISBN-10 981-4307-91-2

Printed in Singapore by World Scientific Printers.

This book is dedicated to my students: those who asked questions as well as those who didn't.

Preface

In *Hindsight* I take a critical look at the work of several important astronomers. My motivation for doing this as well as my methods are spelled out in the first chapter, but before getting to that I want to make some preliminary points.

This is a critical look, which means sometimes (not always) I will have disapproving things to say. It does *not* necessarily follow I think myself superior to the astronomers I criticize. In fact any contribution I may make to astronomy is negligible compared to that of Sir John Herschel, say, or Sir James Jeans. But the character of science is such that I am allowed to discover a mistake in either man's work, bring it to the attention of other scientists and (depending on what it is) have it corrected.

My criticism is based on a very limited part of each astronomer's work, and in particular on one or two books written for a popular audience. I do make comments about things like scientific style and rigor; these must be understood to refer to these works alone, and the same scientist might look very different in writing (for instance) his technical papers.

As a book about books, I use many verbatim citations from my authors. I have tried to be exceedingly careful not to quote out of context, and in general not to give a misleading impression of what any of them actually said. Unavoidably I have omitted sections that were irrelevant to my particular point (shown by ...). Words in square brackets [] are my own, either replacing or amplifying those of the author cited. Any references to page numbers that don't include a specification of a particular book refer to the one being examined in that chapter. Where a page number has "n" behind it, the reference is to a footnote on that page (p. 365n, for example). Other letters appear as Sir John Herschel appended material in later editions of his book, §369a being an update of the original §369.

Another note on dealing with old books: some of the works I'll be referring to were reprinted over many years in several editions, and the versions I have at hand are not always those originally printed. There should be no differences in content, however there is a possibility for confusion as to dates. The first volume of Laplace's *Mécanique Céleste*, for instance, originally came out in 1799, but I am using an English translation printed in 1829. To give the proper chronology of each work, I will refer to it in my text using the originally published date, and in my bibliography will add the date of the version I'm actually working from. The latter one will probably also be easier to get hold of nowadays, for those interested in looking at something first-hand.

Finally, I'm going to ask you to pay attention. I think I've been reasonably successful at avoiding mathematics and at explaining things so no background in science is necessary (though I have put in a few equations for those who can use them), but many of the points I'll be making are subtle and some of the arguments need careful thought. There doesn't seem to be any way around this; the universe can be a complicated place.

Alan B. Whiting

Contents

Chapter 1

One Question and Two Ironies

"All this astronomy stuff is marvellous. But is it true?"

Marvels astronomy certainly has. There are pictures of unearthly beauty (literally), multicoloured clouds and pillars, shining by starlight or some strange light of their own; details of other worlds, with rocks of ice and methane rain; stories of vastness and duration beyond any sane imagining, to the point that astronomers talk about the beginning of time in dead earnest. It has become impossible for poets to use hyperbole in writing of the science, for when they seek to exaggerate, it has already passed them.

And there is a great interest in the science from people who are not astronomers, indeed from non-scientists. On bookstore shelves one finds everything from guides to the night sky for binocular-wielders to expositions of theoretical cosmology, in a section that may not quite match those of Romance or Self-Help in size but reliably holds its own. On a more active level, telescopes open for public viewing are perennially popular; I recall one night when the modest instruments at Cambridge seemed to have emptied southern England. I have worked in the field for decades, but still don't quite understand why there should be so much curiosity among non-astronomers about such an arcane and esoteric study, one with minimal practical application or effect. But so it is.

So there are marvels, and people who have had no hand in them and so might indeed ask how much they should really believe. Normally it's expressed to me in more particular terms: "Are there really black holes?" or "Is the universe actually expanding?" But the overall question is the one that really needs to be addressed: how far can a non-scientist trust what he or she is told by a scientist about science?

In principle, if one assumes the textbook character of a science, this is a meaningless question. You should not need to *believe* at all. Science is

1

supposed to be objective, not authoritative; the answer should not depend on who works it out. If an astronomer presents a result you should be able to look over the figures, check the calculations, if need be proceed from the original raw data all the way to the conclusion, and even duplicate the experiment or observation yourself. In fact that's what astronomers do to each other. Even if they are not inclined to repeat everything they can check important bits of it, and test it against other results and against well-established conclusions about the way the universe works (and the way the science works). This happens even beyond the context of scientific research: in thousands of classrooms worldwide millions of students, most of whom are not notable geniuses, check the work of Newton daily. (They don't see it that way, for the most part, but in a very real sense it's true.)

Non-scientists, however, by definition lack the tools to do that kind of review. They are forced back onto some kind of judgement about the credibility of astronomers, as a whole or in the particular case before them. This is the first irony: most people's knowledge of that essentially non-authoritarian process, science, is based on some kind of scientific authority. So the question of how far you, as someone outside astronomy, can believe what an astronomer says is a real and important one. I hope it's clear that some results are well-established, some less so, so that the question is not, "Should I believe Astronomy?" but "How far should I believe what I am told?" And that "how far" includes both how much of the overall quantity of facts and stories to believe, and how much credence to give each one.

I could approach this question in a very current context. I could select several expositions of astronomy for the public and work out (using my own observations and computer time, if necessary) how far you should believe them. It's still a possible way to work, but I won't do it that way for two main reasons. The first is the practical difficulty of duplicating the work of a great number of very inventive and active people. The second, and more important, reason is that for much of what is being presented I can't give an answer even in principle. Cutting-edge science is, by definition, that area for which we do not have clear answers, and it's the most exciting stuff, the parts that are most in demand by readers. (You will sometimes hear an astronomer answer a question after giving a seminar by saying, "That is a subject of current research." The alternative, "We don't know," is exactly equivalent.) This is the second irony: exciting science may not be true; science known to be true, well-tested and reliable may not be exciting.

Instead, I will take a somewhat indirect approach. Looking at Astronomy as a single thing and taking the position of a lay-person reader, I ask, how much of what authoritative scientists have told us over the years has turned out to be true, and how much untrue? — and then go into some details of why. It's the kind of criterion people use to assign trust among themselves: if someone is known to make statements that are later shown to be wrong, whether by intention or simple mistake, one is cautious about believing any new things they might say. I propose to apply this technique of hindsight to a range of books of popular astronomy, all published long enough ago for us to judge what they got right and wrong.[1]

Now, hindsight has a bad reputation, and most of it is well-deserved. When looking at the past in the light of present knowledge and ideas it is easy to be unfair. One can convict people for violating laws not yet passed and castigate people for not seeing the invisible. (Arthur Koestler does this in *The Sleepwalkers*[2] (Koestler, 1959), seeing Newtonian mechanics as self-evident and blaming a sort of long-lasting, collective psychological disease for the failure of anyone to formulate it in the millennium before Newton.) On a slightly different level it leads to "Whiggish" history, in which everything is seen as contributing to or resisting some later development, that (perhaps implicitly) there is a purpose in history having to do with the later result. Hindsight in science is especially dangerous from this point of view: one can trace, for example, the various roots of the Quantum Theory and how everything finally came together to produce that great edifice. But the apparent goal is only visible in retrospect. At the time nothing of the sort was clear, and to use the goal as an organizing theme is both unfair to everyone involved as well as misleading.

However, used properly I think hindsight can give us, if not the full answer to the question at hand, at least a valuable clue about which way the answer might lie. It is something like the answers to the odd-numbered problems in the back of the textbook, telling us whether the student understands the material or whether a rereading of the chapter and class notes is in order before starting again. By examining situations in which we know the answers, we gain insight into the reliability of the process.

---

[1] How long is long enough to settle a scientific question has been investigated by Virginia Trimble (Trimble, 2008), with answers that may be surprising. As far as possible I will avoid anything about which there are doubts still.

[2] I'm not implying that you should have read this book or that you should necessarily go out and read it. It's here as an example to show that an assertion I've just made is true, so you can if you wish go out and check it yourself, or perhaps use it to follow an idea of your own. This is the purpose of citations in scientific and scholarly literature. Actually, *The Sleepwalkers* is well-written and widely read, so it's a pity it has certain basic flaws.

As a sort of general philosophy, my technique adopts the attitude that several authors writing over the space of a century or so may be lumped together as one thing, Astronomy, to be trusted or not according to a criterion normally applied to a single person. There is certainly some unfairness in such a starting point. But to a lay person a science often does appear to be a single entity, to be dealt with as such.

In combining my lay-person point of view with hindsight and picking out right and wrong statements I of course risk gross misuse of the technique. It would surely be terribly unfair to fault Sir John Herschel, for instance, for failing to foresee the complicated developments in dynamics that led to chaos theory a century after his death, or the results of quantum theory that eventually explained the power source of the Sun. My emphasis must be on what could be known *at the time* about the reliability of a stated conclusion. Along these lines I will develop something of a cautionary list for scientists, a taxonomy of errors, reasons that certain results presented to the public did not turn out to be true. This list may possibly be of some interest to historians of science and more (I think) to current scientists. But, again, it seems to me that someone viewing Astronomy from outside is justified in complaining when he's told something that turns out to be untrue, regardless of what the explanation eventually turns out to be. However interesting or difficult the process is, we are entitled to judge it by the results.

In adopting this approach I will be cutting through other subject-fields and other questions at an angle. That means I will be touching on matters that are important in their own right, and deserve thought and attention, but which I will not address. For instance, I have mentioned the objectivity of science only to show that it does not operate for the layman; I do not intend to investigate to what degree or in what ways astronomy is actually objective. Likewise, belief is an enormous psychological field and certainly how and why people trust each other (or do not) is terribly important to a science as to any human activity. Related to this is authority, how people become authorities in science, what sort of authority they have and what the word means in practice. I will not work with any of these directly. In this book I am developing only one criterion (though, I think, an important and useful one) that may reasonably be used to assess credibility, and I shall avoid dealing with the details of authority by choosing authors whose status as authorities is not in any doubt. If one cannot use Sir John Herschel's work on astronomy as an authoritative exposition, after all, whose can one use?

Since the books were written over the space of a century and more, a period in which astronomy itself was changing greatly, along the way you should get a feel for progress in the science. This is not my main aim. In fact I want to emphasize I am *not* writing history. I am examining a question by taking a historical approach; but to write history itself I would have to include a depth of coverage and treatment and to consider a great number of ideas and concerns that would lead me far from my main question. I want to answer my question, or at least go some way toward the answer, and a full and proper history would only be a distraction. Similarly, I am not doing a review of popular astronomy as such. That would call for a study of many books written by people who were not noted authorities in the science of astronomy, an in-depth study of the intended and actual readership, what the ultimate effect of the books was (if that can be determined), and many other matters that would lead us far from the question I've set out to answer.

Lastly, I will be dealing with a science, and to understand much of what I'll be saying you will have to understand some of the science itself. I have written the book so that no mathematical skill is necessary to understand it (though a little can be very helpful, and I've included some equations for those who can use them), and will be providing some backgound on physics and astronomy as we go along. But it is not my aim to teach you astronomy or mathematics. That, again, would take time and attention that would lead us away from the main question (and one could argue that it would defeat my purpose by making you into scientists — to some degree — and so leaving the lay-person status whose point of view I'm trying to maintain). I will try to provide enough in the way of background and explanations for you to follow the main thrust of the book, while being sufficiently brief and concise that neither of us is much distracted. My background notes will thus be terribly unbalanced: I will leave out some elementary things as unnecessary and include some rather advanced ideas as they come up.

I will be dealing specifically with astronomy. It is only one of the sciences, smaller than most in population and professorships even if it has a long history and a high public profile. It has its own methods and results, many of which don't travel well beyond its borders. But I expect to find some answers and insights that will be applicable, even useful, across Science. In fact astronomy as a science has two features that make it a good subject for this sort of analysis: it has enough mathematical and physical rigor to allow wrong answers to be identified unambiguously (if not always right ones, and not always immediately); and it has been rigorous for long enough to allow a historical approach like this one to be attempted.

To lay out my plan in detail: below is a list of several books on astronomy, written by undoubted authorities in the field with an intended audience of non-scientists. For each of these books I will start by sounding like a book review. This is not for the purpose of getting you to buy old books (driving up their prices in secondhand bookshops!); it is necessary for the proper exercise of hindsight. One must have an appreciation of the context in which the book was written, what it contains and what its stated aims were. Sir John Herschel (I'll look at two of his books in detail) admonishes astronomers to know their instruments "to extract from their indications, as far as possible, all that is true, and reject all that is erroneous" (Herschel (1833), p. 67, §103). Context is one way of checking on the operation of this troublesome instrument, hindsight.

Having set the stage, my plan is to look at what each writer got right; what he (alas, I have no female authors) got wrong; whether he could or should have had suspicions about the latter; and whether the reader could have picked up any clues about the matter independently. This leads into an investigation the taxonomy of errors, which will develop as we go along.

The books I'll be examining are:

- Sir John Herschel's *Treatise on Astronomy* from 1833 as well as a late edition of *Outlines of Astronomy* from 1869, the period between being one of great development in the science;
- Sir George Biddell Airy's *Popular Astronomy* from 1868, by an Astronomer Royal;
- Simon Newcomb's *Popular Astronomy* from 1878 and the much later *Astronomy for Everybody* from 1902;
- Sir Robert S. Ball's *In the High Heavens* from 1893, by a Professor of Astronomy at the University of Cambridge;
- Sir Arthur Stanley Eddington's *Stars and Atoms* from 1928, and a contrasting view of very much the same material,
- Sir James Jeans' *The Universe Around Us*, appearing in two quite different editions in 1929 and 1944.

So keeping in mind that we will be dealing with a science, but not studying it; dealing with just one science, but with an eye toward Science as a whole; taking a historical approach, but not studying History; and working out credibility without studying Belief, let us look as what some brilliant men got right and wrong.

Chapter 2

Positions, Orbits and Calculations

Before we go into what someone has said about astronomy we need to know a few things about the science. If you're already familiar with it, well and good; you need not pay close attention (at least, until we get to some of the more advanced ideas). This chapter is intended to give someone with no background a few of the terms and ideas that underlie nineteenth-century astronomy.

It is not intended in any way as a complete or balanced introduction to the science. There are important elementary things I'll leave out, and some pretty advanced ideas I need to mention. My initial purpose is for you to have at hand certain pictures and ideas, so I can (metaphorically) point to them later on when I need to. Secondly, when the discussion turns to how (and how well) something was being done, it helps to know in some detail what the astronomers were in fact trying to do. Elias Loomis, introducing his 1855 book *An Introduction to popular Astronomy* (which is rather more a technical manual than a popular astronomy book in our sense), pointed out "...no one can feel a rational confidence in the results announced by astronomers without some distinct notion of the methods by which those resuts are obtained" (Loomis (1855), p. v). At the same time, Sir John Herschel warns that reasoning on the basis of an analogy or description of the mathematics, rather than actually doing the math, can lead to incorrect understanding (Herschel (1833), p. 331, §523n), so be aware that what I am about to try has its limitations.

For the purposes of the next few chapters the major task of astronomy can be summarized overall as *observing* the *positions* of celestial objects, that is using *visual* observations made by looking through *telescopes*, and then *calculating* where those objects will be in the future.

To start out with, consider that astronomy deals very much with angles.

You're familiar with how they're measured: there are 360° in a full circle, 90° in a right angle, and you can probably estimate (especially if you've had practice) the measure of an angle drawn on a piece of paper. In astronomy we often deal with much smaller angles (and sometimes specify large angles to high accuracy), so we also use minutes (60' to the degree) and seconds (60" to the minute). These are an example of the sheer skill of astronomers at using confusing words (it won't be the last); these seconds and minutes have nothing directly to do with time. Both sets of words have a common origin in the Babylonian *sexagesimal* system, based on counting in sixties. Nowadays we use mostly decimal systems, but the sexagesimal system has its advantages: the half, third, fourth, fifth, sixth, tenth, twelfth (and so on) part of a degree are each a whole number of minutes. Minutes and seconds as measures of angles are sometimes called "of arc" to make them unambigious. To give you a feel for the sizes of angles we'll be dealing with, in his book of 1833 Herschel characterizes 1' as a "gross amount" and points out that 1" is "distinctly measurable" (Herschel (1833), p. 65, §102) with astronomical instruments, a statement that holds more or less throughout the period of the books I'll be looking at. A dime seen at a distance of roughly two miles (about three kilometers) subtends a second of arc. This kind of comparison may impress you that telescopes can see very small things; but is not terribly useful for astronomical purposes, in which there are few dimes an nothing is as close as two miles. I will have further comments on the use authors make of numbers as the matter comes up.

2.1 Looking at the sky

What do you see when you look into a clear sky? Leaving aside the Sun and the Moon for the moment, you see the stars. These are points of light, of no discernible size and of widely different brightnesses. The ancient Greeks ranked them as first magnitude (the brightest stars), second, and so forth down to the faintest that they could see, the sixth magnitude. When instruments were devised to measure brightness objectively, it was found that the average star of the sixth magnitude is about a hundred times fainter than the average first magnitude star. This terminology and relation still define the way astronomers describe brightness.

The stars are far away, farther than the trees or the mountains on the horizon, though it's not immediately clear how much farther; all you can tell for sure about their position is a direction. If you watch for many minutes you will notice that those in the east are slowly getting higher while those in the west are getting lower. If you watch carefully for a longer time you can see that those in one part of the sky (the north, for those of us in the northern hemisphere) aren't changing their height much, but instead circling around a particular direction. Eventually, if you stick at it, you'll find that their motions behave as if they are all attached to a single enormous sphere, centered on you, rotating once a day (*diurnal motion*) around an axis oriented north and south. This is the *celestial sphere*, and it's in that elite class of extremely useful objects that don't really exist. Though it's not actually there, it is often convenient to use as a short of verbal shorthand and talk about things "on the sky" as if they really are attached to this great hollow globe. Not marked by anything on the sky, but obvious to you, is the point directly overhead, the *zenith*.

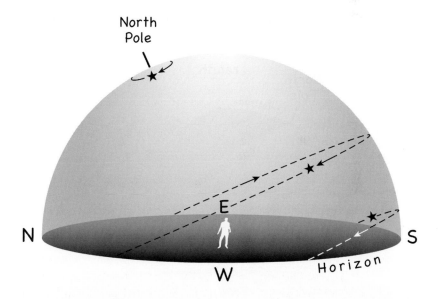

Fig. 2.1 How it appears to an observer. The seemingly flat Earth extends to the horizon in all directions evenly, forming a circle. Objects in the sky move as if they were attached to the inside of a great hollow sphere, centered on the observer, which rotates on an axis pointing North-South and inclined to the flat Earth at an angle that depends on the observer's geographical location. In this drawing the observer must be in Alaska, or the very north of Britain, since the North Celestial Pole is very high in the sky. Notice also that geographic north, the north marked on maps, is along the ground, not up in the sky, and so is a *different* north. It is the *projection* of celestial North onto the ground, as close as you can get to the celestial North direction without flying.

In fact it's also useful mathematically. When dealing with directions, one can draw and solve a *spherical triangle* on its (nonexistent) surface. By "solving" a spherical triangle we mean using the measurements of some parts (lengths of sides or possibly angles between the sides) to calculate other parts. An astronomer might, for instance, measure the angular distance of a star from his horizon (giving the length of one side of a spherical triangle, the *zenith distance*) at a given place (yielding another length, the *colatitude*) at a certain time (giving the angle at one corner, the *hour angle*), then determine the star's distance from the north celestial pole (the length of the last side). The formulae to use involve things like sines and cosines, as one uses in triangles on a flat surface, but are rather more complicated.

Of course, this idea has to be used with care. Since we are actually working with pure directions the celestial sphere must be treated as if it were infinite in size, or the Earth just a point; that is, when a star is directly overhead it's not really any closer to us, or farther away, than when it's on the horizon. And this idea of the stars being attached to a

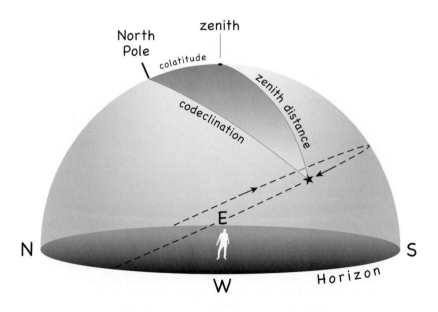

Fig. 2.2   A spherical triangle used to solve a problem in positional astronomy. The vertices are the North Celestial Pole, around which the celestial sphere appears to rotate; the zenith, directly overhead from the observer; and the position of the star in question. Knowing or measuring two of the sides and one of the angles allows the astronomer to calculate the last side.

hollow globe implies that they all keep the same patterns of directions. That is, although they're all revolving around us once a day, in relation to each other they're not moving. This is an idea that will need to be modified later on, but it's convenient sometimes to talk of the "fixed stars." As well, the patterns themselves, the *constellations*, are convenient to use as a short way of specifying a general direction in the sky.

Fixed stars imply moving stars, and there are a few of these. Some of the brighter ones, while taking part in the general diurnal motion, don't quite match it exactly; they move among the fixed stars at various rates. One of these *planets*, called Saturn, takes about thirty years to move around the sky (relative to the fixed stars). The Sun takes a year, and the Moon a month, so these were anciently numbered among the planets (bringing the total to seven).

Among the fixed stars are a few patches of cloudy light, not brilliant points like stars are. They are not terribly exciting to look at with the naked eye (except possibly if you count the Milky Way among them) and don't figure prominently in astronomy before the age of telescopes, but "fuzzy stars" were anciently noted, for instance, in the constellations Orion and Andromeda. With the coming of telescopes more of these *nebulae* were found, some of which were resolved into clouds of stars (like the *globular cluster* of stars in Fig. 2.3). But not all.

If you are very, very careful and patient about watching the sky (or if someone has alerted you to the possibility) you may notice that a few stars change their brightness over time. The idea was so strange to people used to the fixed stars staying unchanged for centuries that the first one to come to the general attention of astronomers was called Mira, "the wonderful." It goes from being a medium-bright star to so faint as to be invisible to the naked eye, and back, about every 330 days. There is a different star (called Algol) which dips in brightness very regularly every 2.86 days. With the coming of telescopes more of these *variable stars* were found, also.

Before telescopes the only bodies in the sky to show any features were the Sun and Moon. Of course, watching the Sun is a difficult and hazardous business; even with the naked eye you can destroy or seriously damage your eyesight, and people have done so. But with the proper filtration and especially with the magnification that telescopes provide, one can often see *sunspots*. These are dark areas, continually changing in size and shape and lasting for a couple of weeks or (rarely) over a month. They show that the Sun rotates, but not as a solid body, since they go at different speeds at different latitudes. I suppose you would liken them most closely to clouds, not permanent and always changing but still definite *things* at any moment.

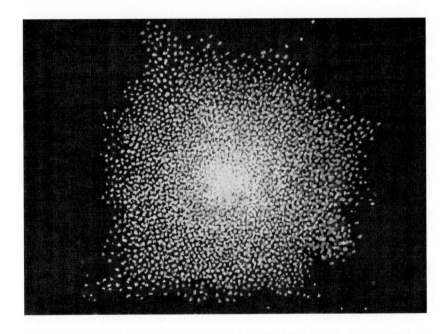

Fig. 2.3 A drawing of a globular cluster, from the book of Sir John Herschel of 1833. In a small telescope it would look like a fuzzy patch of light. Naturally this raises the question whether *all* nebulae are in fact clouds of stars like this, some just too far away to resolve.

The Moon has light and dark markings visible to the unaided eye. With even the smallest telescopes *craters* become prominent, of all sizes down to the smallest detail that your telescope will see. They are round rings of mountainous material surrounding a lower, sometimes saucer-shaped area; often there is a central mountain. Some look older and more worn, some younger and sharper. Some are clearly newer than others, overlying them and modifying their shapes. All in all, the Moon shows the most detail to an Earthbound observer of all the things in the sky, appearing indeed like another world.

The main instrument used by astronomers in these chapters (and indeed, in some form, to this day) is the *telescope*, an arrangement of lenses and possibly mirrors whose details we need not go into here. For our purposes they have two good characteristics: the main, big lens or mirror, the *objective*, gathers light, more light than our eyes take in alone, so with a telescope we can see fainter objects. (That means that stars fainter than the sixth magnitude appear, and one can talk of seventh, tenth, or even fifteenth-magnitude stars; remembering that the higher number means less light). And, by selecting the *eyepiece* lenses properly, we can view a magnified image of the sky. The amount of magnification is technically limited only by the steep curves one must use for some lenses and the size of the telescope overall. In practice, the diameter of the objective sets a limit on how small the details are that one can see (for reasons having to do with the nature of light), and often one is limited to even less magnification by the unsteadiness of the atmosphere (called *seeing*). It is a rare night and a rare telescope that will deliver resolution much smaller than a second of arc.

Looking through a telescope one can see an image of the sky and the objects in it. But often a scientist will want to measure something, and the image is not something you can really put your hands on or take a ruler to. For this one can mount in the telescope a thin wire or hair, a *crosshair*, and arrange it to be in focus at the same time as the image, superimposed on it. Then when a star is on the crosshair the telescope is pointed at the star very exactly. To measure small distances in the telescope field of view another crosshair can be mounted and allowed to move, connected to a screw whose turns measure how far it's gone. This is the *bifilar micrometer*, sometimes just called the micrometer. Though there are other types of micrometer found in use, this is the most popular one.

A number of other new and modified aspects of the sky become visible once you have a telescope. There are more nebulae and variable stars, as I've noted. Then the previously starlike planets are shown to have details: Mercury and Venus show phases like the Moon does (crescent, half, gibbous), although no clear craters or other surface marking appear. Mars has bright caps at the poles and surface markings which appear to be permanent, though details change. Jupiter and Saturn have bands of dark and light colors and ephemeral spots. The latter has a flat ring around it, the former four bright satellites (eventually Mars, Saturn, Uranus and Neptune are also found to have satellites, and more are found for Jupiter). The fixed stars, on the other hand, still show no surface features, seeming pointlike even in the best of telescopes under the best conditions. But something else very important appears: some are binary stars, two (or more!) stars appearing to revolve around each other in the same way that planets revolve around the Sun, or satellites around most of the planets. Their orbits as observed in telescopes are generally very long, dozens of years or more, but the binary nature is clear.

So, incidentally, "fixed stars" are not completely fixed relative to each other. But orbits taking many years and measuring seconds of arc are small corrections to the concept of the celestial sphere. Stars also have small motions apart from binary orbits, *proper motions* amounting to at most seconds of arc per year, again a small correction to the idea of fixed stars.

2.2   Making maps

If you have a suspicion that something changed in the sky, you won't really be sure unless you know what it looked like beforehand — that is, unless you have made a good map. A star map is also useful if you're interested in the detailed motions of celestial objects, as nineteenth-century astronomers (in particular) were, or even if you want to be able to find the same thing again. To make an accurate star map is a process of endless careful tedium.

The main instrument for this operation (for the purposes of this chapter) is something called a *transit telescope*, or sometimes a *transit circle*. You want a telescope, of course, because it allows you to magnify what you see, and thereby distinguish between directions (positions on the sky) that are very close to each other, giving a better and more accurate map. It is mounted solidly so that it can only swing north and south along one line, the line passing directly overhead. On it is a scale showing the angle along this line, which you can use to find a star's distance from the north celestial pole (in the northern hemisphere, of course). To find the position of the star in the other direction, east or west, is harder because of course it's moving all the time with the diurnal motion. So instead of measuring an angle we use that motion, and measure the time. We have a well-made, well-regulated clock inside the room with the transit telescope and note the time at which the star just passes the central cross-hair in the eyepiece, and refer it to the time of some particular thing we've chosen as a reference.

Step back a moment and look at what we're doing (in Fig. 2.4). The transit circle is using the axis of the Earth as a reference direction: as the Earth rotates, only the places where this axis meets the sky (the celestial poles) are stationary, that is as far as diurnal motion is concerned. And as the Earth rotates it carries the transit circle around, pointing at various parts of the sky at various times. Our map of the sky is thus inevitably based on features of the Earth. In fact the coordinates used to describe where an object is on the sky are taken with reference to the Earth, one

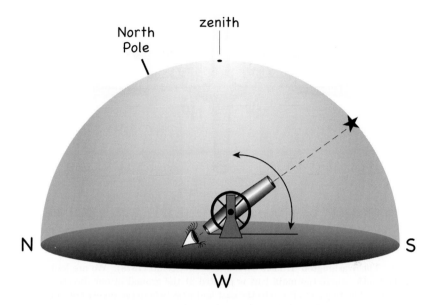

Fig. 2.4 The principle of the transit circle. Here it's shown from the west side, mounted on the apparent flat Earth, with the celestial sphere rotating overhead. It swings at the axis of the wheel, rotating only in one line; this is the line that takes it through the celestial pole and the zenith, and observes objects just as they pass this line.

showing how far it is from the pole and the other how far from the *equinox*, a position related to the Earth's orbit.

Some of those features, unfortunately, are not quite as simple as I've explained so far. For instance, the axis of the Earth changes its direction with time. It wanders around a circle on the sky; sometimes it's better visualized as a cone traced out by the axis itself. The diameter of the circle is about 47°, so it's not a small effect overall; but it takes 26,000 years to go completely around. This is the *precession of the equinoxes*, and it means you have to specify a reference date for your star positions (as well as keeping close track of the time in order to find those positions in the first place). On top of this big cone is a smaller wobble, called *nutation*, with a period of about 19 years. So after a long night of carefully measuring the angles and times for a number of stars, the astronomer must stay up for much of the day to take a pile of raw observations and apply corrections for precession and nutation — and several other things.

One of the most annoying of these is the Earth's atmosphere. Any starlight, or planet-light, comes to a telescope after passing through dozens

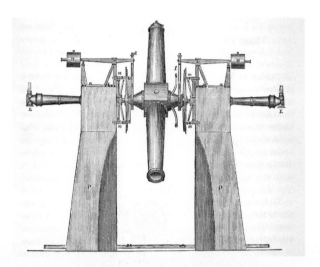

Fig. 2.5 The Washington Transit Circle, from Newcomb (1878). We are looking at it from the south, where the main lens is pointed at the ground in our direction, maybe for cleaning. The two piers P are to the east and west, where the pivots are. The wheels where the angle from the pole is read are edge-on to us.

or hundreds of miles of this. Effects that pass unnoticed in normal life can be troubling on this journey. One of these is refraction, in which light coming through the atmosphere at an angle (that is, from any object that is not directly overhead) gets bent downward, so that a star appears higher in the sky than it actually is. Another is *seeing*, in which chunks of air of different temperatures and pressures refract the light differently, so that light coming from a single direction appears to come from several. The pointlike image of a star gets spread out into a blob, and details on the surface of a planet are smeared together. (To get an idea of what seeing is like, look just along the top of a road on a hot day. Anything you see beyond the road will be distorted. Or you could find a telescope, even a small one, and look at a star under high magnification.)

To apply the exactly proper correction for refraction one needs to know the temperature and pressure of the air all along the path that the starlight travels — which is quite impossible in practice, so there always remains a little uncertainty about the exact measurement. In addition there is the problem of finding the exact position of the star when you're looking at a blob enlarged by bad seeing. These contribute to the *probable error* or the *uncertainty* of the measurement, along with residual irregularities in your angle scale (how to you know all your degrees are the same size?), slight misalignments of your transit circle, play in your gears, changes in the speed of your clock with temperature, and many other things. The word "error" in this context is unfortunate, since it implies that someone has done something wrong, which isn't the case; it has misled generations of science students. "Uncertainty" is better. Working out your uncertainty in a given measurement is as important as the measurement itself — and generally more difficult. The point is that, if you say the position is *here* to within a tenth of a second of arc, and it is calculated to be four seconds away, something is wrong. On the other hand, if you're looking for an effect that only makes a difference of a hundredth of a second, you just won't see it in this observation.

So if you have a given instrument and have everything adjusted as best you can, you will still have some uncertainty in your measurement. To reduce that, the popular method is to make multiple observations of the same thing (assuming your target is such that you can). In this way, the random uncertainties will tend to cancel each other out, in the long run, and you get a more accurate answer. You do have to be careful to be truly random (measuring in the morning and evening, all around the sky, in summer and winter, etc.) or you may introduce a *systematic error*, which is a really pernicious beast.

If you're measuring something periodic, a very effective method is to measure as many periods as you can. If you can fix the time a certain feature on the face of a planet crosses the center to within five minutes, say, and time three hundred rotations, you've measured the time of one rotation to within one second. (This is how the ancient astronomers, who could only measure the time of something to within a few hours — a "watch of the night" — could come up with the length of the year to within a few seconds: just measure over hundreds of years.)

2.3   Kepler's orbits

Let's shift from observation to the basic picture derived from those observations. For the purposes of this chapter we take the Solar System (all those objects whose motions are dominated by the gravity of the Sun) to be described, for a first approximation, by Kepler's laws of planetary motion. The Sun is found in the middle (though not exactly at the center of each orbit). That is, each planet orbits the Sun, its path over time being an ellipse with the Sun at one focus. It changes its speed according to a definite formula, speeding up when closer to the Sun and slowing down when farther away. The overall time a given planet takes to go around the Sun depends on the size of its orbit, also according to a definite formula.

An ellipse is a specific type of closed curve, a sort of flattened circle. The degree of flattening is measured by a number called the eccentricity; a circle is an ellipse with zero eccentricity. The eccentricity increases with flattening until it reaches one, which actually denotes a parabola (the ellipse has been flattened so much that it's broken at one end). An ellipse with an eccentricity of one-half is sort of moderately flattened.

A line from one side of the ellipse to the other, the longest way, is the major axis; half of its length is the semi-major axis, a number used to describe the size of the ellipse. The two foci of each ellipse lie on the major axis. For a circle they sit on top of each other in the center (or, if you will, there is only one focus); as the eccentricity increases they separate, heading for the ends of the ever-more-flattened orbit, though even for a parabola the focus stays a measurable distance from the curve itself. For the Solar System, the Sun occupies one of the foci (and nothing special happens at the other). The points on an orbit closest to and farthest from the Sun, which is where the major axis hits the curve itself, are respectively the *perihelion* and *aphelion*.

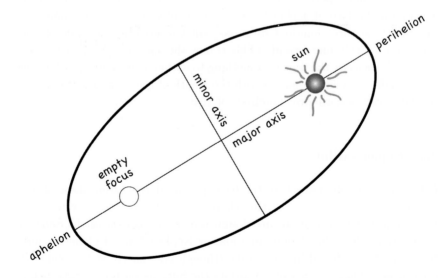

Fig. 2.6 An ellipse, the prototype for planetary orbits, with parts labeled.

For a given planet once we have the semi-major axis (size) and eccentricity (shape) of the orbit, and put the Sun at one focus, we have a good picture. But an ellipse is a plane figure and the world is three-dimensional. The plane of one planet's orbit is not the same as another's. Taking the Earth's orbit as the reference (since we make all our measurements while standing on it), the *inclination* of a planet's orbit is the angle between its plane and Earth's. There are other numbers used to describe the orientation of an orbit, the *argument of the perihelion* and *longitude of the ascending node*; and of course we need to specify a time, usually given as the time of perihelion. For our purposes, though, only the eccentricity, semi-major axis and inclination need to be kept at hand.

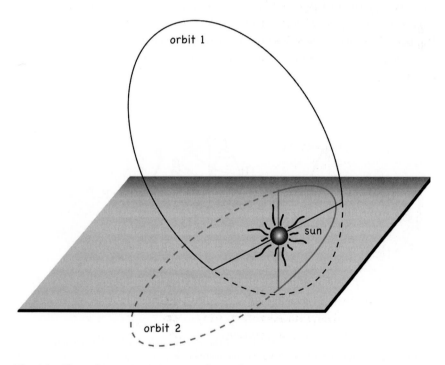

Fig. 2.7 Two orbits around the same object, for instance two planets around the Sun. The Sun is at one focus of each orbit, but the sizes and inclinations of the orbits need not be the same. Here the planet with the smaller orbit spends most of its time below our reference plane, while the outer one has most of its orbit above the plane. No major planet in our Solar System has such highly inclined and eccentric orbits as these, though they could be taken as representative of some asteroids or comets.

The picture I've just given is specific to the Solar System. With a few changes of nomenclature and details, it applies to the motions of satellites around planets and even to the orbits of distant stars around each other.

One last subject needs to be mentioned before we leave orbits. The axis of rotation of a planet points in the same direction (with small and slow changes, like precession and nutation) regardless of where it is in its orbit around the Sun. If it's not exactly perpendicular to the plane of its orbit (and the Earth's is tipped by about 23°), that means one pole will point toward the Sun for part of the orbit and away for the other part. Indeed, one whole hemisphere (northern or southern) will be pointed more toward the Sun for part of the orbit, and less for the other part. As seen from the planet, the Sun will be higher in the sky and spend more time above the horizon for part of the orbit, and be correspondingly lower and spend less time for the other part. This is the cause of the seasons (see Fig. 2.8). They can be modified somewhat by the eccentricity of the orbit itself, as the planet is carried closer to and farther from the Sun, but in the Earth's case the eccentricity is small and does not have much effect.

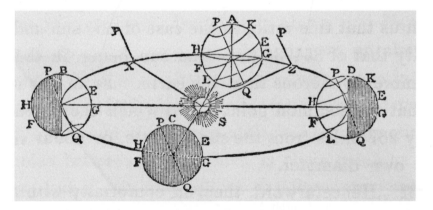

Fig. 2.8   The seasons come from the fact of the Earth's axis not being perpendicular to its orbit, so that for part of the orbit each hemisphere is tipped more toward the Sun and for part of the orbit each is tipped more away from the Sun. Here, P is the North Pole and Q the South Pole. (We need not worry about all the other letters.) The diagram is wildly out of scale: the Sun is actually much larger than the Earth, and the diameter of Earth's orbit is over a hundred times the diameter of the Sun. Beware of diagrams like this! They are meant to illustrate one idea only, and can be very misleading otherwise. This drawing is from Herschel (1869), and is identical with that used in Herschel (1833).

2.4 Distances

I began this chapter by emphasizing that at the outset we do not have distances to astronomical objects, only directions. Once we have the picture of Keplerian orbits, though, that's certainly not true for planets. The relative sizes of the orbits are fairly straightforward to work out, though it's a bit tedious to be precise. Even the absolute size of the System as a whole can be figured (with a little less accuracy). What about other objects? For comets and asteroids, small (if interesting) members of the Solar System, we can employ the same methods used with the planets. For stars we need to use the whole System–or at least that part of it we can visit. We make use of the size of the Earth's orbit and something called parallax.

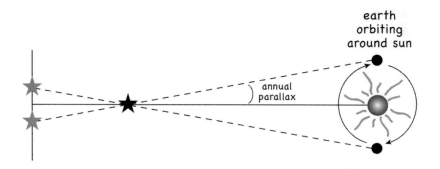

Fig. 2.9 Finding a star's distance by parallax: on the right is the Earth's orbit, in the center the target star, on the left where the star appears to be when compared to even more distant objects. As the Earth moves in its orbit the direction to the star changes, and one can use trigonometry to measure its distance in terms of the size of the orbit. The figure is not at all drawn to scale. No star is close enough to show an annual parallax as large as one second of arc.

Looking at Fig. 2.9, notice that the direction from the Earth to a star on one side of the orbit (say, about September) is different from that on the other side (March). For very distant stars (which is all of them) the difference isn't much; the angle there at the sharp tip is the annual parallax of the star, always less than a second of arc, and gives its distance more or less directly (the smaller the angle, the farther away). Normally one tries to measure the angle, not by determining the star's position with a transit circle, but by measuring its change of direction against more distant stars in the same telescope field. It's still a difficult and delicate measurement, and as the next chapter opens had not actually been done.

2.5   Calculations

The basis of all calculations mentioned in the next few chapters is Newtonian mechanics, the system of working out motions first set out in the
Principia Mathematica well over a century before. The actual solving of a
problem in this way requires the use of calculus (which Newton invented
for the purpose); I will try to give you a description and a picture of what
is going on mathematically.

Each little bit of stuff in the universe — technically, each pointlike mass
— has associated with it something called (linear) *momentum*. It's probably
best to construct it in three steps. First, take the direction the object is
heading, say north-northwest. Draw an arrow in the right direction. Next
adjust the length of the arrow to show its speed, longer for faster; at this
point the arrow represents the object's *velocity*. Finally, adjust the arrow
again to be longer if the stuff is heavier, more massive, and shorter if it's
lighter; now it represents the object's *momentum*. (Mass is not the same
thing as weight, but for our purposes you can think of it as a quality that
can be measured by how much the object weighs.) A light object going
fast can have the same momentum as a heavy object going slowly, but two
identical objects going the same speed in different directions have different
momenta.

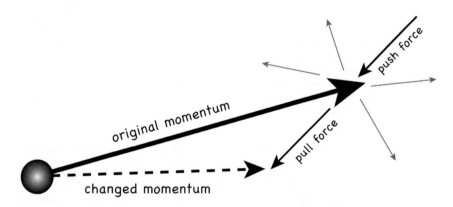

Fig. 2.10   A momentum-arrow, showing all at once where an object is going, how fast
it's going, and how much mass it has, being affected by a force-arrow. The force can
act in any direction, so it can speed up or slow down the object, or perhaps change its
direction without affecting how fast it's going. You can think of a force as either pushing
on the end of the momentum-arrow or pulling, changing the momentum-arrow into a
different one. The light arrows show other possible directions the force might have acted.

An object's momentum can be changed only if you apply a *force*. In a sort of circular definition, a force is defined as something that changes momentum, and a momentum can only be changed by a force. Now, you can change momentum by speeding the object up, slowing it down, or making it turn — changing its direction. So think of force as an arrow also, tugging on the end of the momentum-arrow, whose effect depends on which direction it's pointing.

Gravity is one of the several possible forces you can use in a Newtonian calculation. Its force-arrows all point inward toward a bit of mass, getting longer (stronger) as you get closer. To work out how something moves as it's being affected by the gravity of another object, you start with a certain momentum at a certain point; adjust the momentum according to the gravity force there; move forward, according to the new momentum, to the *very next point* (this is important); adjust the momentum according to the gravity at this point; and so on. (Or you can work backwards if you know its path, to give the gravity.)

Look at Fig. 2.11 and think of how the body moves coming in from the right. If it were coming straight in, its momentum would only increase in size without changing direction, and it would speed up until it hit the mass in the centre. Coming in at an angle upward, its direction is changed to the left and its speed increased. The change in direction and speed would get greater as it got closer to the centre, then fade away as it got farther.

Now consider what would happen to a body that has some size, not just a point as we've been thinking so far. If you put it in the gravity field of Fig. 2.11, that part of it closer to the mass would feel a stronger force than the part of it farther away. It would tend to get pulled apart along the direction to the mass (and slightly compressed at right angles to that line). This is the *tidal force*, which gives rise to the ocean tides on Earth. It shows up in other astronomical contexts, too, and is important whenever bodies get very close to each other. In particular, if something like a planet's satellite gets too close to the planet, it can get torn apart by tidal force. (This is one idea for the origin of Saturn's rings.)

Going back to the basic problem of finding the motion of a body, Newton's method of working out motion only really applies in an infinitesimal region: it starts as a *differential* equation. If the force or momentum is changing, you have to refigure everything *at every point*. It says nothing (directly) about what happens any distance away — as opposed to, say, a Kepler orbit, in which you know you'll be somewhere on the orbit forever. In this aspect it's a very limited tool. On the other hand, you can in prin-

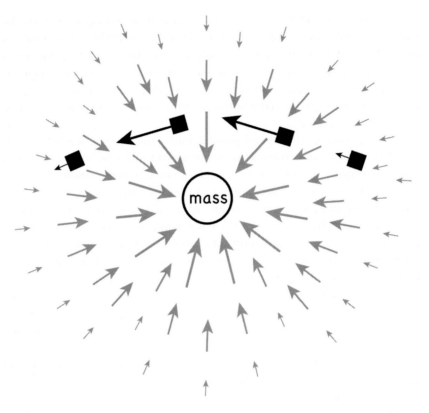

Fig. 2.11   A representation of the gravity exerted by a mass, showing a field of force-arrows. Every point in space in principle has a force-arrow, but only a few are shown, for clarity in the figure. Here a body enters from the right, speeding up as it approaches the mass and slowing down as it moves away, having its path bent all the time.

ciple work out the path of an object under any kind of force, even a very complicated one or one changing all the time. "Solving" the differential equation means finding this path. You are trying to match the force-arrow with the change in the momentum-arrow; when you can do that, you've found the solution. There are three ways to do this.

First, if you have a simple or very symmetric situation, you may find a solution given by an expression you can write down, a **closed-form solution**. (This is an equation which gives you the position for each moment of time, so you get an equation as a solution to the differential equation.) For two mass-points moving only under their mutual gravity you do find good Kepler orbits as possible solutions. If you allow them to be spheres of some size, instead of strict points, you get the same answer. If you make the spheres flattened, or add a third mass-point, suddenly you have no closed-form solution. (Sometimes you can find special situations in which a closed-form equation results, like one in which all objects fall together into a centre, but that's not likely to be applicable to the physical situation you have to deal with.) You could even decide that your situation is close enough to a closed-form situation that you can, with a certain error, use it even though it's not strictly applicable.

Second, you can find an approximate solution by setting up the field of gravity arrows and putting in a momentum arrow, figuring out the changes, and stepping forward a small but not infinitesimal bit in time (and space); then running through it all again. If you do this you're assuming that things are changing slowly enough that you're not making much of an error. You are moving in a series of short straight lines instead of a smooth curve, and you're hoping that you get close to the same place in the end. In fact it can be a very complicated matter to figure out how close you actually come this way, and to do it right takes an enormous amount of calculating and recalculating. This approach was not popular until the advent of fast digital computers. But it can give you an answer in situations of *any* complexity.

Last, you can decide that the situation you want to calculate is close to one of the simple situations, and start with that. Suppose you have a planet in orbit around the Sun, with another planet in its own orbit. You set up your Kepler orbit, but this time the gravity-arrow does not quite match the change in the momentum-arrow because of the gravitational force from the other planet. You figure out how much the difference is, and adjust the path accordingly. In the new path the gravity-arrow does not quite match the change in momentum-arrow either, since it was figured for the gravity *on the old path*; but if you've chosen your situation well the match will be much closer, and your answer will be closer to the right one. Then, if you want to be more accurate, you can figure out the gravity-momentum mismatch on the second path, and work out a third path. All this assumes that, as you make small changes to the path, you get correspondingly small changes in the gravity-arrows.

It is this method of **perturbations** that was used for hundreds of years in working out motions of objects in the Solar System. It is still very calculation-intensive, but not as much as the second method. And it gives very good results — most of the time.

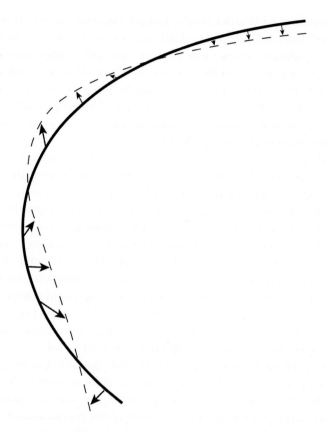

Fig. 2.12 A picture of working out an orbit by the method of perturbations. Most of the force-arrows come from the gravity of the Sun, so the initial guess as to the planet's orbit is a Keplerian ellipse (dark line). Along this ellipse the mismatch between the force and the change in momentum is figured out (arrows), and the orbit is changed in the right direction to correct this. The corrected orbit (dotted line) only makes the change in momentum match the gravity force along the *old* orbit, however, so it doesn't quite fit. If it's not close enough, you can go through the procedure again.

There is another, more strictly mathematical way to look at the method of perturbations. In the differential equation you've written down and are trying to solve, there is some term, some expression, the symbol representing some physical thing, that you just can't work with. It resists all attempts to transform it into a nice form that you can manipulate and solve. Well, you replace it by a sum of things that you can work with, generally an infinite sum. You have to assume (and justify it when and as you can) that the first term in the sum is *most* of the sum, and the second term takes up *most* of what's left, and so on. It becomes a tricky thing, sometimes, even to show that an infinite sum of your terms actually adds up to something finite. The process of perturbations corresponds to starting with the first term in this sum, working things through, and then picking up the second term, and so on. Often you have more than one set of terms, and you work with the first one of each; then fold in the second one of each; and so on. (Sometimes the perturbation is in something rather abstract, not easy to draw, maybe not possible to link directly with a drawing. The procedure is the same.)

The assumption behind the perturbation method, whether you think of it in a picture or as a series of mathematical terms, is that a small difference in your input produces a small difference in your output. That is, moving a small distance away from one point will make a small difference in the force of gravity there, and in the path you need to follow to match up the force-arrow and the change in momentum-arrow.

There are more than two bodies in the Solar System, so no planet actually follows a Keplerian orbit. A planet will not return exactly to the same place after revolving around the Sun once; its actual path is a pretty complicated one. However, it is convenient to consider it as following a Keplerian orbit that is changing with time. For *this* instant we can work out what the eccentricity, semi-major axis, longitude of perihelion and all would be; and then consider that it is actually on this orbit; and that these numbers are changing slowly. This gives the orbit itself a sort of reality, even though the planet (strictly speaking) doesn't follow it for more than an instant. But it's another useful nonexistent thing, along with the celestial sphere.

As I've outlined it, Newtonian mechanics has been applied mostly to point masses and spheres. It can actually be extended to solid bodies of any shape, as well as fluids (liquids and gases) and many more complicated situations. In fact both the precession and nutation that I've mentioned in the motion of the Earth can be calculated by Newtonian physics. And one can calculate motions in a more abstract way, so that the arrows and the paths represent more than one body's trajectory through familiar three-dimensional space, or a body's trajectory through more than spatial co-ordinates. The details can get terribly complicated, but you're still using one of the three methods to find a path with the right relationship between momentum-arrows and force-arrows. Eventually, when dealing with other situations giving rise to differential equations, like the run of temperature and pressure inside a star, you can use the same sorts of ideas and methods.

2.5.1   *Conservation laws*

Although Newtonian physics is initially presented in terms of figuring forces and changes in momenta on a very small (differential) scale, there are some ideas that apply in the large. They can be used to work out certain characteristics of a situation without going through all the details, sometimes giving you all you want to know.

First, there is the *conservation of linear momentum*. Force changes momentum; but if you put a bunch of objects in a box, or otherwise keep outside influences out, the only forces present will be among themselves. Then one finds that, if you add up all the momenta, you will always get the same number (and direction of arrow). The objects, whether they are planets or pool balls, can change each other's momenta but the sum total stays the same.

Second, there is the *conservation of angular momentum*. Angular momentum is built up out of rotations and the distribution of mass in a rotating body, the way linear momentum is built up of motion and simple mass. You can have a light body rotating rapidly have the same angular momentum as a heavy body rotating slowly. You can also have a big, spread-out body rotating slowly but having the same angular momentum as something else with the same mass all concentrated toward the center, rotating quickly. The classic illustration of this is an ice skater, spinning faster as she draws her arms in toward her body. Her mass and angular momentum stay the same, but by concentrating her mass toward her center the rotation rate must compensate by going up.

In astronomy the most important example for our pupose is a gaseous blob, something that may eventually become a star or planet. It may not be spinning perceptibly at the beginning, but as it contracts under the action of gravity its mass gets more concentrated, and unless it has exactly zero angular momentum (which is hard to arrange) it will develop a definite rotation.

Angular momentum is changed by a *torque*, the way linear momentum is changed by a force. A torque is made up of force and the point at which it's applied; a force that acts directly through the axis of rotation exerts no torque. In astronomy toques are generally far weaker than forces, but become important over the long run.

The third conservation law is that of energy. A discussion of that will be postponed for a few chapters, until we're ready to look at thermodynamics.

2.6   Stability

A rather advanced concept, at least if you're actually going to do calculations, is *stability*. The simpler and easier type of stability is static or equilibrium stability.

An object is in *equilibrium* when all the forces on it (there can be lots of forces, of all different types) cancel each other out. A push one way is met by an equal push in the opposite direction. That means the momentum-arrow isn't changing, and in fact could be zero (which means the object is not moving at all). If you hold a ball steady in your hand, gravity is balanced by your pull, and the ball is in equilibrium (leaving aside the motion of a spinning Earth for the moment). A planet in a circular orbit is *not* in equilibrium: there is only the force of gravity, unbalanced, and its momentum is changing all the time (in direction, though its speed stays the same).

An object in equilibrium is *stable*, or has *positive stability*, when any small push away from equilibrium generates forces that send it back. A pyramid sitting on its base on a table is stable to any kind of small over-turning force; if you pull it one way, it will fall back. An object in equilibrium has *neutral stability* when a small push generates no (different) forces. A cylinder lying on its side on a table is neutrally stable. And an object is *unstable*, or has *negative stability*, when a small push generates forces that move it farther from equilibrium. A pyramid balanced on its tip on a table is unstable.

These examples have only one, or maybe two, directions in which the object can move (whether the motion is stable or not). Often things get more complicated. A violin string, for example, can move in an infinite number of ways, corresponding to the various notes it can sound; a drum-head has two dimensions of infinite numbers of motions. But it turns out (very usefully) that any motion can be decomposed into a set of standard motions (called *normal modes*), and the stability of each one worked out. If any of the modes are unstable, the whole situation is unstable. It's like putting a marble on a saddle: the marble is stable to forward and backward motion, where the saddle bends up, but unstable to sideways motion, where the saddle bends down. The marble will fall off eventually, so it's unstable overall.

We'd like to extend this intuitive concept of stability to a more dynamic situation. Is, for instance, an orbit stable? It's not in equilibrium, so the previous definition doesn't apply. In fact we can define dynamic stability in different ways. We could say it's stable if, after a small disturbance, it eventually comes back and intersects the undisturbed orbit. We could say it's stable if, after a small disturbance, it never gets more than a small distance ("small" here is a well-defined mathematical idea) from the undisturbed orbit. Or we could define it as stable if, after a small disturbance, it eventually comes back to within a small distance of the undisturbed orbit. These are different ideas and have different mathematical implications. (Think of a two-dimensional pendulum. If we just pull it to one side and release it, it demonstrates the first kind of stability. If we pull it aside and give it a sideways push, it will demonstrate the second kind, but not the first. Of course, a pendulum at dead center is in equilibrium, so the analogy with a dynamic system is not exact.)

When we deal with more than one object in an orbit things get even more complicated. Is the Solar System stable if the planets all stay in the same order; if none ever gets flung out to the cold, interstellar regions or falls into the Sun; or if all the semi-major axes and eccentricities stay within a small region of their present values? The answer one gets depend on the exact form of the question one asks.

In this chapter I have gone through a great deal of material quickly, some of it very advanced. Don't worry if you do not grasp everything at once! Be ready to refer back to it, as we begin our survey with an exposition of the state of astronomy in the first third of the nineteenth century.

Chapter 3

Sir John Herschel, *Treatise on Astronomy*, 1833

The year is 1833. Britannia rules the waves, Andrew Jackson is serving as the seventh President of the United States, and an increasing amount of coal is being burned to fuel the Industrial Revolution. No steamship has yet crossed the Atlantic (under steam), though, and most people have never traveled faster than a galloping horse.

Physics as formulated by Isaac Newton has been generally accepted as the truth by "natural philosophers" for over a century, though applications and implications are still being worked out. Front-line research telescopes have lenses approaching a foot in diameter and at their eyepieces are small number of professional astronomers and a large number of amateurs. There are larger telescopes, using mirrors instead of lenses, but the manufacture and use of these tricky speculum-metal instruments seems to be almost a family monopoly of the Herschels. One of the latter has just brought out a book on astronomy for the public.

Sir John Herschel was born to the science. His father, William, an immigrant Hanoverian, had developed the art of making reflector telescopes to a degree not surpassed for decades, and then used his creations to survey the sky diligently and carefully. Along the way he discovered Uranus; not really by chance, since his completeness and diligence ensured that he could not miss anything of the sort. His work on the contents and distribution of the sky, in Admiral Smyth's[1] phrase "celestial architecture," was eventually of rather more importance to astronomy; but the planet earned him a royal pension and leisure to pursue his work (current astronomers may muse

---

[1] Admiral William H. Smyth's *Cycle of Celestial Objects* is the forefather of the plethora of observing guides for amateur astronomers now available. Even today it has some claim to shelf space for anyone who looks through a small telescope at the sky. It also shows a great deal of learning and is engagingly written. Unfortunately its audience is rather too advanced for my purposes, so I have not included it in this book.

upon the difference between what is effective at getting funding and what is scientifically important). As his book comes out Sir John has already been active in astronomy and this year will begin to extend his father's survey to the southern sky during a long visit to the Cape of Good Hope. He will be a central figure in astronomy generally during the mid-nineteenth century and will contribute as well to physics and photography. A book about astronomy with the Herschel name has authority. This one is also popular, judging from the fact that later versions would appear decades hence (one of which I'll look at in detail).

That does not necessarily mean that the book is well-written, clear, or in any way useful for someone seeking to learn about the science. "Astronomy by Herschel" could easily have been something merely to lend status to one's bookcase; great scientists are not necessarily good at describing their work to the public at large. And, indeed, a modern reader opening the volume could be forgiven a sinking feeling: the first sentence takes up half the first page. This is nineteenth-century prose indeed, endless phrases of qualification and extension, great multisyllabic learned-sounding words.

The sinking feeling should be resisted. For the proper operation of hindsight the writing must be taken on its own terms. If the modern reader can maintain concentration through a long, meandering sentence and can develop the patience to sit through an extended discussion, there is much rewarding material here.

(One administrative note: Herschel broke up his book into chapters and sections, the latter being a useful way of keeping track of things when a book is printed in editions with different page numberings. I will provide both page and section numbers when I refer to a particular part.)

3.1   The purpose and the readership

According to Sir John the time is ripe for an exposition of astronomy, now (a third of the way through the nineteenth century) a mature science. Such an exposition is "Practicable, because there is now no danger of any revolution in astronomy, like those which are daily changing the features of the less advanced sciences, supervening, to destroy all our hypotheses, and throw our statements into confusion." (p. 3, §4) No danger of revolution in astronomy! Any historian could count at least three major revolutions in the science since this book's time and possibly argue a few more. But note the qualifying phrases: no revolution *like* those in other sciences; there is no danger of *all* our hypotheses being destroyed. Herschel leaves open the possibility of some parts of the science being changed, possibly changed greatly. He asserts, in effect, that there is a solid part of astronomy that has been established for all time. It's a claim that actually belongs to the field of the Philosophy of Science, and perhaps not strictly in the path of our investigation here; I'll return to it briefly in the last chapter. For now, notice that the long and ramified nineteenth-century sentence is just the sort of thing to use if you want to make precise statements. Herschel had the right tools at hand to use when he had something complicated to say.

Within this mature science Sir John's aim is "simply to *teach* what we know," (p. 3, §4) as opposed to trying to prove the correctness of its conclusions or to convince a skeptic to abandon contrary beliefs. This takes the book out of the sometimes strident Newtonian-anti Newtonian polemic of the seventeenth and eighteenth centuries, which indeed had not quite finished at the time of this *Treatise*. At the same time, "...our aim is not to offer to the public a technical treatise ..." (p. 8, §10) suitable for turning the reader into a research astronomer. Herschel considers that he is doing something original, that "...it is something new to abandon the road of mathematical demonstration in the treatment of subjects susceptible of it, and teach any considerable branch of science entirely or chiefly by the way of illustration and familiar parallels ..." (p. 6, §8) So he felt himself to be at the beginning of what we might call popular astronomy. Of course, to do astronomy right one needs "...*a sound and sufficient knowledge of mathematics, the great instrument of all exact enquiry, without which no man can ever make such advances in this or any other of the higher departments of science, as can entitle him to form an independent opinion on any subject of discussion within their range.*" (his italics; p. 5, §7) So to form one's own conclusions about this kind of science, that is to exercise the objective character of the field and get away from a reliance on authority, one needs to be at home with math. To explain things without mathematics will be sometimes inordinately difficult, but he's going to try.

His method is to *describe* the mathematical process which gives a certain result, "that central thread of common sense on which the pearls of analytical research are invariably hung" (p. 8, §10), then to go on to "illustration and familiar parallels," that is, example and analogy. By showing something in as many ways as possible, different styles of learning may be accommodated (a sentiment that has a very modern ring). The picture thus given may even be useful for an actual astronomer familiar with all the relevant mathematics; Sir John's work may "...convert all his symbols into real pictures, and infuse an animated meaning into what was before a lifeless succession of words and signs." (p. 7, §9) The use of several different aspects to build up a coherent picture of a given part of astronomy, as we shall see, is one of Herschel's important themes.

Who was Sir John writing for?

What he means by "the public" is not the present mass market for books. In 1833 a large fraction of the British population was quite illiterate, and the majority simply could not read a book of this level. He assumes that his readers are familiar with the Copernican theory, enough that he need not defend it; and that they also have a knowledge of "...decimal and sexagesimal arithmetic; some moderate acquaintance with geometry and trigonometry, both plane and spherical; the elementary principles of mechanics; and enough of optics to understand the construction and use of the telescope, and some other of the simpler instruments." (p. 4, §6) He is serious about this. There are several rather intricate geometric proofs (for example, p. 49, §69, and pp. 115-6, §175); he expects his readers to know what a versed sine is without being told, and he trots out spherical triangles fully expecting the reader to know how they are solved; sections of the *Principia* are referred to where appropriate. In addition, in two places he constructs analogies of planetary motion (p. 268, §420, and pp. 361-2, §569) that depend on the reader being at least vaguely familiar with the mathematics of two-dimensional pendulums.

On a slightly different level, Sir John uses several Latin quotations, though he gives translations of them all[2] (translations are not always found in astronomy works even decades later), as well as Greek etymologies (also translated) of some scientific terms. He has a classical background, but his audience is not made up of classical scholars.

So the readership is made up of literate people with a fairly high level of skill in basic and secondary mathematics, though they are not mathematical scientists themselves. It is definitely a minority of the population of Britain at the time, but large enough for Herschel to call it the "public." To write science for them is, according to the author, a new thing. It will be interesting to see how later authorities thought of their readers.

In modern terms the readership would probably correspond to those who had a bachelor's degree in one of the mathematical sciences, along with those of the less numerate public who were willing to skim over the algebra and trigonometry and take the results on trust.

---

[2]Except one or two, like that on p. 13, §16, which he seems to have overlooked.

3.2 Content

The book is divided into thirteen chapters and an introduction (from which I've already quoted freely), arranged so:

Chapter and contents	Number of pages
Introduction	9
I. General Notions	61
[the Earth: form, shape and size]	
II. Of the Nature of Astronomical Instruments	43
[details of construction and use]	
III. Of Geography	50
[surveying and mapping]	
IV. Of Uranography	27
[mapping the sky]	
V. Of the Sun's Motion	29
[motion, size and shape]	
VI. Of the Moon	19
[motions, size and shape]	
VII. Of Terrestrial Gravity	11
[also solar and lunar gravity]	
VIII. Of the Solar System	45
[planets: motion, size and shape]	
IX. Of the Satellites	12
[motion, size and shape]	
X. Of Comets	12
[orbits and tails]	
XI. Of Perturbations	60
[calculation of departures from pure elliptical orbits]	
XII. Of Sidereal Astronomy	36
[stars and nebulae]	
XIII. The Calendar	8
[briefly, the basics]	

Note that the second longest chapter, on the details of calculating accurate orbits, deals solely with fine details of the motions of planets. These are the *planetary perturbations*, the deviations from strictly Keplerian orbits caused by the gravitaional pulls of other planets. They are calcuated using the equation-solving method of perturbations, as outlined in the last chapter, but are formally a different thing. (It should be evident from the context which kind of "perturbation" I'm talking about in any given place.) Likewise, most of the chapters devoted to celestial bodies deal with their location and movements; material on the Earth is included explicitly because it is the base from which observations of position are made, and other major sections deal with the instruments that make those observations and the theory of gravity used to calculate all motions. This is early nineteenth-century astronomy: positions and motions, what we now call astrometry. As Sir John himself says, "The great object of astronomy is the determination of the laws of the celestial motions, and their reference to their proximate or remote causes." (p. 77, §121)

First comes the great mapping, done with incredible patience and attention to detail by visually examining the sky with telescope and cross-hair to the ticking of a finely-adjusted clock; then the tedious corrections and calculations done by hand (though the invention of log tables makes the work much easier, p. 263, §416). Herschel does allow that astrophysics, that is the "constitution and physical condition, so far as they can be known to us," (p. 9, §11) of celestial bodies is a part of the science of astronomy. But the caveat is important. Before the application of the spectroscope to astronomy physical observations are very limited, and before the development of the science of thermodynamics the physical interpretation of observations is sparse and uncertain (as we shall see).

There is a little in the way of physical astronomy, mostly descriptive. After summarizing observations of sunspots Herschel asks, "But what *are* the spots? Many fanciful notions have been broached on this subject, but only one seems to have any degree of physical probability ... " that is that they are holes in the luminous atmosphere of the Sun allowing us to see the opaque body below (p. 208, §332). Note that this explanation is presented with the faintest possible support: it is the only thing anyone has thought of so far that is not ruled out by sheer absurdity. As to the source of the Sun's energy, we are "completely at a loss" (p. 212, §337).

Of the planets, they are certainly opaque bodies shining by reflected sunlight; some have atmospheres, some do not, about some we are not sure. Mars, for instance, "...has been surmised to have a very extensive atmosphere, but on no sufficient or even plausible grounds." (p. 279, §437n) Comets (beyond their orbits, which have been calculated reliably), "...are as much unknown as ever. No rational or even plausible account has yet been rendered of ...their tails ..." (p. 300, §470). It appears probable that departures from calculated comet orbits over a long period of time, together with the observed characteristics of the tails, indicate that the Solar System has a thin sort of atmosphere resisting cometary motion; but the fact of a resisting medium has not been established, and raises many questions (p. 310, §484n).

Moving out to the universe beyond the Solar System, we have one short chapter. (This material would form the bulk of any modern book on astronomy, at least any not devoted specifically to the planets.) About the distance, size and intrinsic brightness of stars "... we know nothing, or next to nothing ..." (p. 374, §584). No star is so close as to have a parallax of one second of arc, so the distance of the nearest is larger than 19,200,000,000,000 miles; "How much larger it may be, we know not" (p. 378, §589). Binary stars, one of the most important of the areas of work Sir John inherited from his father, show that Newtonian gravity appears to operate well beyond the Solar System (p. 391, §606). There are stars that change their brightness, some periodically, some apparently not, but so far in spite of many suggestions there is no sure explanation why (pp. 380-5, §593-7). Some nebulae are certainly made up of stars, though possibly not all; it is an ongoing argument (more on this particular discussion below).

To summarize: the overwhelming bulk of the book, reflecting the astronomy of the time, is concerning with locating the various celestial bodies and working out their motions through the theory of Newtonian gravity. A few limited conclusions regarding the physical nature of things out there can be worked out, but most of what we would call physical astronomy remains a mass of doubtful observation and speculation, noted as such.

3.3 Themes

There are three important themes (and one minor one) running through the book that I would like to highlight. The first is the great difference between the apparent and the real, between what the sky looks like at first glance and what long, involved research has shown it to be. This is indeed a very conscious motif of Herschel's from the very beginning, for in his introduction he notes that "Almost all [astronomy's] conclusions stand in open and striking contradiction with those of superficial and vulgar observation and with what appears to every one, until he has understood and weighed the proofs to the contrary, the most positive evidence of his senses" (p. 2, §2). This is still a major point of learning and surprise to students, but nowadays the problem is almost the reverse of what Herschel identified. Now students are used to getting most of their information from digested sources, TV and Internet and whatnot (still, to some degree, from books), so the constructed real picture is familiar; when they first go out under the sky the apparent view can come as a shock. In Sir John's time, before these sources of information and before electric lights, his readers could be assumed to be familiar with the appearance of the sky but possibly not with the full constructed picture behind it.

So the (apparently) flat Earth is carefully shown to be round, by the phenomena of a ship going hull down and only the peaks of mountains being seen from far away (pp. 15–16, §20-1), as well as by the shape of Earth's shadow during lunar eclipses (p. 220, §349); and because the constellations do not change size or shape between rising and passing overhead, the Earth must be immeasurably smaller than the distance to the stars (pp. 49–50, §71-2). Well, this is hardly cutting-edge stuff, even as long ago as 1833; in fact these same phenomena were noted, with the same explanation, seventeen centuries previously by Ptolemy, and weren't new then. A bit further on, the subtle and mathematical processes of precession and nutation are noted as being further proof of the correctness of the Copernican (heliocentric) picture of the Solar System (p. 173); this in spite of the fact that in the introduction Sir John has declared he will *assume* Copernicus in all that follows. Again we are dealing with old material. Why all this unnecessary and archaic demonstration?

It is all in accordance with a second theme, Herschel's declared aim of showing his picture from a multitude of viewpoints with the purpose of building up a coherent whole. Rather than depend upon a single "decisive experiment," or even a few key observations, he is showing how everything hangs together by looking at the theory from all possible sides. In fact this may be a more subtly powerful idea than he himself realized. Scientific theories are not always proved or accepted by a single clear experiment, but often by the accumulation of marginal results in many instances. The idea of a "decisive experiment" may have an important use in pedagogy, but it has much less in the actual conduct of research.

This leads directly to the third theme: a succession of increasingly accurate theories and observations, called by Herschel in another place (Herschel (1869), p. 769, §856) residual phenomena. Herschel's own words are probably the best to explain the idea:

> The steps by which we arrive at the laws of natural phenomena, and especially those which depend for their verification on numerical determinations, are necessarily successive. Gross results and palpable laws are arrived at by rude observations with coarse instruments, or without any instruments at all; and these are corrected and refined upon by nicer scrutiny with more

delicate means. In the progress of this, subordinate laws are brought into view, which modify both the verbal statement and numerical results of those which first offered themselves to our notice; and when these are traced out, and reduced to certainty, others, again, subordinate to them, make their appearance, and become subjects of further enquiry. Now, it invariably happens (and the reason is evident) that the first glimpse we catch of such subordinate laws — the first form in which they are dimly shadowed out to our minds — is that of *errors*. We perceive a discordance between what we *expect*, and what we *find*. The first occurrence of such a discordance we attribute to accident. It happens again and again; and we begin to suspect our instruments. We then enquire, to what amount of error their determinations can, *by possibility*, be liable. If their *limit of possible error* exceed the observed deviation, we at once condemn the instrument, and set about improving its construction or adjustments. Still the same deviations occur, and so far from being palliated, are more marked and better defined than before. We are now sure that we are on the traces of a law of nature, and we pursue it till we have reduced it to a definite statement, and verified it by repeated observation, under every variety of circumstances. (Herschel's italics; p. 70, §111)

In this process there is no decisive experiment (as noted above), just an accumulation of results not quite in accordance with theory. Examples are given, among others, in the case of the refinement of the "solar motion" (pp. 75 and 206, §118 and 329). Hence the great importance of knowing one's instruments, the stated rationale for the whole of Chapter II; leading to, for instance, the details of how one measures an angle with a finely-graduated transit circle (pp. 82-3, §128), including the operations of clamping and unclamping the scale. And in general Herschel pays much attention to the quantitative level of uncertainty in all measurements as a limit on how far one can test the theory, often giving numbers when a more modern book might be content to say "very small." It's worth pointing out that this attention to the numerical size of uncertainties is something on which enormous time and effort is spent in every modern-day science class (with equivocal results).

So Herschel is concerned to present his picture of astronomy from as many angles as possible, and is acutely aware of the *quantitative* limits on what he can say.

There is one minor theme I want to mention. It is the near-certainty of there being intelligent life elsewhere in the universe:

> ...he must have studied astronomy to little purpose, who can suppose man to be the only object of his Creator's care, or who does not see in the vast and wonderful apparatus around us provision for other races of animated beings (p. 380, §592).

Although this particular passage concerns other stars, comments are made in many places referring to possible inhabitants; for instance, about how the universe might look to a Saturnian (p. 286, §446), cautioning the reader that what to a human might seem dark and dreary could in reality be far otherwise and a "benificent contrivance."

In no case does this conviction lead Herschel into error. He does not assert that any particular extraterrestrial forms of life actually exist, though some of his speculation seems at best premature. Indeed, it could be argued that it is beneficial, in stimulating the imagination into a more careful and detailed consideration of places and situations strange and distant. Other authors were not always so restrained.

3.4   Hindsight

Now turning the pitiless glare of hindsight on Sir John Herschel, what do we find?

3.4.1   *Tentative results and weak support*

There is nothing to fault in the sections where he frankly says he doesn't know what's happening. As already mentioned, this includes the nature of comets (especially their tails), what makes variable stars change brightness and the source of the Sun's energy. He also does not know what light is (p. 180, §282n). And even though the motion of bodies under the influence of gravity is something he understands (as we'll get to soon), he is at a loss to apply his analytical techniques to the many-body problem of a globular star cluster (p. 400, §615).

In other areas he presents tentative results and interpretations, identified as such. We've already noted that his idea of sunspots is described with the weakest possible recommendation, and the fact of Mars' atmosphere is far from proven. In a slightly different vein he presents a picture of the water-driven weathering of Earth's continents not as a formal idea, but just to show that there *could* be a mechanism to wear them down (pp. 120-1, §182-3). (In fact water erosion is an important factor in Earth's surface geology, but the picture presented by Herschel is not really the way things work. We'll get back to this.)

He also points out that he cannot rule out the existence of liquid water somewhere on the Moon, even though he's convinced that it has no atmosphere (pp. 229–230, §364). In this he is correct; though we know *now* that water cannot exist as a liquid at such low pressure,[3] the state of the science of thermodynamics (or physical chemistry) in 1833 was such that Herschel did not know.

Why present such tentative and uncertain things at all? It appears that Herschel is aware of the need for imagination in the initial stages of formulating anything new in science, and as long as speculations or "dreams" are identified as such they are harmless (p. 277, §434). This is an idea that we'll see again in other forms, stronger when we reach the twentieth century.

3.4.2 *Good insights*

Herschel must be credited with many good insights, especially considering the undeveloped nature of astrophysics in his time. His experience as an observer shows in his dismissal of the "Moon Illusion" as just that (p. 34, §47), as well as his scepticism about reports of the apparent projection of stars onto the face of the Moon (pp. 220-1, §349n). The appearance of double stars in complementary colors is more due to the structure of the human eye and brain than to the stars themselves (pp. 394-5, §610). And the structure of the Orion nebula is too diffuse and visual observation too uncertain to credit reports of it changing shape (p. 403, §619).

One could almost call prophetic his statement that there are probably more planets than were known (pp. 243-4, §387), and he believes that "...the number of stars may be really infinite, in the only sense in which we can assign a meaning to the word." (p. 373, §582)

I believe he also deserves some credit for his caution concerning the nature of the nebulae. This is a subject worth some comment, for it becomes important later on. He, his father and his aunt had spent years finding and cataloguing thousands of these fuzzy patches of light. William had constructed a theory in which some kind of self-luminous matter, a sort of fluid, was originally spread out thinly and evenly; then coalesced under the action of gravity, forming more and more well-defined and brighter nebulae, eventually becoming stars and star clusters (summarized in Herschel's later book, which we'll get to in Chap. 6, and more fully in Newcomb's first book, which we'll look at in Chap. 7). With each advance in telescope power and refinement more nebulae had been resolved into stars, raising the question

---

[3]Water *has* been found on the Moon, but in the form of ice, probably bound in some way to the soil, not liquid at all.

of whether this self-luminous fluid actually existed, or was only the product of William's imagination and inadequate observations of distant objects. Sir John summarized his view of the situation this way:

> The nebulae furnish, in every point of view, an inexhaustible field of speculation and conjecture. That by far the larger share of them consist of stars there can be little doubt ... On the other hand, if it be true, as, to say the least, it seems extremely probable, that a phosphorescent or self-luminous matter also exists ... what, we ask, is the nature and destination of this nebulous matter? (pp. 406-7, §625)

His conclusion that there probably is this strange stuff comes from the way certain nebulae *appear* in the telescope.

> The nebulous character of these objects, at least of the former [the Great Nebula in Orion], is very different from what might be supposed to arise from the congregation of an immense collection of small stars. It is formed of little flocky masses, like wisps of cloud... (p. 408, §619)

Visual appearances in a telescope are not a very firm foundation on which to rest a theory, so the nature of nebulae is still a subject of "speculation and conjecture," and Sir John leaves the matter open.

That clouds would seriously interfere with someone observing the Earth from space and trying to make out the continents (p. 231, §368) is, I think, something not widely realized before the Space Age (refer to any run of science-fiction magazines before about 1960).

One minor point, which is not important yet but will have some significance in a later chapter, is also worth pointing out. The force of the Sun's gravity upon the Moon is greater than that of the Earth's on the Moon (pp. 289-90, §453). Indeed, if the Earth were to disappear, the Moon would not fly off into space, but rather remain in a Solar orbit much as it is now (in fact tracing out a rather simpler path). The matter of orbits within orbits is something we will return to later.

3.5   Wrong answers

Unfortunately, not everything Herschel had to say was right or hedged about with warnings. He is straightforward in his claim that lunar craters "offer, in short, in its highest perfection, the true *volcanic* character ..." (his italics, p. 229, §363). There is no sign of doubt. (He also notes that, although there is no sign of water, some areas are flat and "apparently of a decidedly alluvial character," in the next passage but here his qualification saves him.)

Well, the question of the character of the craters of the Moon lasted almost until men had gone there in person. Herschel is in good company. But it's still true that these are impact craters, not volcanoes. How could a reader have at least built up some doubt?

Consider the evidence on which the statement is based: lunar craters *look* like volcanoes. They do. But many things look like something else; this is a weak sort of support, even though it can appear convincing, especially when visual impressions are almost the only information you have to deal with.

And there is no sign of a competing theory, any other *possible* explanation. Any scientist should be uneasy when the current explanation is the only one anyone has thought of; the alternative may not be right, but the process of comparison and decision makes for a much better theory — and keeps science honest.

The Earth does have volcanoes, of course. One can say, strictly speaking, that they come under geology and thus have no place in a book on astronomy. But Herschel mentions them also, and uses his picture of water-based weathering (which he seems to have forgotten was only an idea to show a possibility) to trace their activity to the Sun (pp. 211-2, §336). For the moment let us extend the earlier caveat and give him the benefit of the doubt, though we'll come back to this.

The problems so far have to do with physics, admittedly the weak part of the book (and the science at the time), and with impressions and stories. The next trouble comes in a calculation and moves partly into gravitational dynamics, the strong point of Herschel's science. Having set out the reason for Earth's seasons, he goes on to show that the varying distance of our planet from the Sun can have nothing to do with them. The amount of heat absorbed by the Earth during the northern winter, when it is in fact closer to the Sun, is just the same as during the northern summer, because it is moving faster in its orbit when it is closer; in fact any way you cut the orbit in half through the Sun, both halves receive the same amount of heat (you might want to refer again to Fig. 2.6). Hence the distance of a planet from the Sun has no effect on its temperature (pp. 197-9, §315).

The calculation, as it stands, is correct. The amount of heat received in any bisection of an orbit *in space* is the same. Herschel's problem lies in the relationship between heat received and *time*. If you give an object a lot of heat quickly the result will be different from giving it the same amount of heat slowly. During the northern winter the Earth is receiving more heat per second, and so should in fact become a bit warmer overall (though effects are completely swamped by the effects of land distribution and weather patterns; the corresponding effect on Mars is much more pronounced).

We cannot fault Herschel for failing to do a proper thermodynamical calculation. As thermodynamics stood in 1833 the relationship between heat and temperature was not at all clear, and a proper, rigorous analysis was simply not yet possible. But I can fault Herschel for failing to understand temperature and heat on a level required for competently grilling a steak, and being misled instead by an elegant mathematical result.

3.5.1   *Gravitational dynamics*

We will leave geology and thermodynamics and now proceed to the strong suit of astronomy in 1833: gravitational dynamics. There is no doubt that this is the center of the science and its pride; Herschel boasts that

> ...at this day, there is not a single perturbation, great or small, which observation has ever detected, which has not been traced up to its origin in the mutual gravitation of the parts of our [Solar] System, and been minutely accounted for, in its numerical amount and value, by strict calculation on Newton's principles (p. 314, §490).

So what in this presentation of Newtonian gravity did Herschel get wrong?

In one very restricted sense, everything. Newtonian physics and gravity have been superseded as accurate descriptions of the universe by quantum mechanics and General Relativity. Planets do not move on orbits in absolute three-dimensional Euclidean space subject to a force called gravity; either (according to Einstein) they follow geodesics on a four-dimensional pseudo-Riemannian manifold in which gravity is curvature, or (in quantum language) wave functions one can identify with planets propagate in such a way, that, when subjected to an observation, their locations may be specified to a given level of probability.[4] Newtonian physics is simply wrong as a picture. (Indeed, the conflicting language of quantum theory and relativity imply that neither of them is wholly correct either.)

But Herschel did not claim absolute truth. He said that, *to the accuracy of available observations,* Newtonian calculations accounted for everything. Indeed, he does not know what gravity is, "be it a virtue lodged in the sun

---

[4]For an explanation of what a geodesic is, as well as the rest of this generally unintelligible sentence, see the later chapter on relativity and quantum mechanics. For now all that's important is that later physical theories, more accurate in describing the universe than Newton mechanics, have a very different set of concepts and structures.

as its receptacle, or be it pressure from without, or the resultant of many pressures or sollicitations [sic] of unknown fluids, magnetic or electric ethers, or impulses ..." (p. 266,§419), only how it acts; it might not even be an attractive force due to the Sun (p. 269, §421). He has deliberately left open the possibility of something like a relativistic description of gravity, even though what Einstein eventually came up with was far beyond anything comtemplated in 1833.

The accuracy of observations also allows him to state truthfully that the rotation of the Earth is constant and that the poles maintain the same geographic location (pp. 40-1 and 171-2, §56 and 266); in fact the day is growing longer and the poles wander a bit, but both effects are too small for the science of 1833 to detect. In the realm of planetary motions and observations, the orbit of Mercury shows a subtle relativistic effect and starlight is bent slightly by the presence of the Sun, but again the instruments and techniques of Sir John's time had not detected these things, and might not have been able to (reliably). Herschel is operating well within his uncertainties.

And these are by no means rough calculations based on inaccurate observations. It is worthwhile to compare Sir John's table of planetary orbits (p. 416) with a corresponding modern set, in Karttunen et al. (1996). (The latter is an introductory textbook for astronomers, and of a significantly more technical nature than Herschel's book, but the details of planetary orbits are not generally given in current popular astronomy books.) Of all the parameters given by Herschel, most agree to within the accuracy listed in the current book. There is a bit of a problem with two angles, the longitude of the ascending node and the longitude of the perihelion, until one realizes that these have actually changed in the interval between Herschel's epoch (1801) and the latter one (1993).[5] With that taken into account, Herschel's values, as Sir John would say, may be used "without serious error."

Herschel's numbers for the sizes of the planets are less reliable, differing by a few percent from current values. I do not think this would surprise him; Mars (5% error) is small and difficult to measure, while Jupiter (5%) and Saturn (4%) are fuzzy by nature, and all numbers in miles or kilometers

---

[5] The figures for the longitude of the perihelion for Jupiter and Uranus disagree by between two and three degrees between the modern figures and those of Herschel, which in this context seems like an unaccepatably large amount. But this angle indicates the direction of maximum diameter for orbits that are almost circular, and so the difference of a matter of degrees is actually rather small.

depend on the distance-scale of the Solar System. (The diameters of the other planets are more accurate). His values of the masses of the planets are excellent for the larger ones (Jupiter, Saturn) and the Earth; but are off by a large fraction for small Mars and distant Uranus, and pretty bad for Mercury. (He doesn't list any masses for the four largest asteroids, since all astronomy could do at the time was place upper limits.) Since all were measured by their gravitational pull on each other, inaccuracy for small objects is only to be expected.

So Herschel's orbits as of 1833 are of high enough quality to use for most purposes now and his planetary diameters are reasonable, not a bad performance at all. His masses need to be updated, but he would be the first to acknowledge that.

There are, however, two matters of dynamics in which Herschel has gone astray. Both have to do with stability.

3.5.1.1 *Stability calculations*

The first concerns the ring, or rings, of Saturn (whether the singular or plural is appropriate is not entirely clear; visually there are at least two, but calcuations mostly consider only one). What are they? "That the ring is a solid opake [sic] substance is shown by its throwing its shadow on the body of the planet, on the side nearest the sun, and on the other side receiving that of the body ..." (p. 282, §441). However, the apparently solid ring raises some dynamical difficulties. Herschel mentions "recent micrometrical measurements of extreme delicacy" that show Saturn's rings to be offset, not centered exactly on the center of the planet, and as a dynamical result "the center of gravity of the rings oscillates round that of [Saturn] describing a very minute orbit, probably under laws of much complexity." This is good, for otherwise they would be unstable, and "the slightest external power would ... [precipitate] them, *unbroken*, on the surface of the planet." (his italics; pp. 284-5, §444). But there appears to be a bit of uncertainty on Herschel's part, for he has not definitely concluded that the stability of the rings is due to their being eccentric,[6] because that would be a result of their being thicker or denser in some places than in others, "of which, after all, we have no proof." As a separate mechanism he says, "we think we perceive,

---

[6] Here I've used the word "eccentric" to mean that the rings are off to one side, not centered on the planet itself. That meaning should not be confused with the eccentricity of a planet's orbit, which is a different thing. I'll only use the word in this way when talking about Saturn's rings, then I'll go back to the orbit meaning.

in the periodicity of all the causes of disturbance, a sufficient guarantee of [the ring's] preservation." That is, because the forces disturbing the ring's equilibrium come from the gravitational tuges of various moons of Saturn and are thus periodic, the situation is similar to "the mode in which a practised hand will sustain a long pole in a perpendicular postion resting on the finger by a continual and almost imperceptible variation of the point of support" (p. 285, §444).

The dynamical system made up of Saturn's (many) rings is a terribly complicated one in detail, and about which we still have a great deal to learn. But, taking things over all and as viewed one-third of the way into the nineteenth century, Herschel's presentation of the situation has real problems. Four of them are apparent. First, there is the assumption that the rings are solid; second, the calculation that an eccentric ring would be stable; third, the observation showing the rings to be eccentric; last, the assertion that a periodic disturbance would stabilize an otherwise unstable ring. I will take each in turn.

Herschel asserts the solidity of the ring because it casts and receives shadows. Well, to be blunt, so does smoke and so do clouds, neither of which can be treated as solid. It's not that no one had considered any other options: the Italian Gian Domenico Cassini had expressed a different opinion a century and more before, and in a work that Herschel was certainly familiar with Laplace (as we shall see) preferred the idea of a sort of fluid. The failure even to consider an alternative (if only to reject it) is a fault in Herschel's presentation if in nothing else. This is apparently an example of a very difficult problem to deal with, that of the unexamined assumption, which we'll return to a bit later.

Next we come to the calculation showing that an eccentric ring is necessary for stability, that is, the ring must be off-center in order to avoid crashing down onto the planet. This appears to have its origin in what is agreed to be one of the great works of mathematical physics.

At the turn of the nineteenth century the great mathematician Pierre Simon, Marquis de la Place, set out the techniques of Newtonian analysis of planetary motions (and much else) in his *Mécanique Céleste*[7] (Laplace, 1799). In Book III, Chapter VI, §46, Laplace considers the stability of a uniform solid hoop under the gravitational pull of a planet. If it is exactly

---

[7]Conveniently for many of us it was translated into English and provided with an extensive commentary (roughly doubling its size) by Nathaniel Bowditch, the first volume appearing in 1829. This is the version I have consulted, though Herschel almost certainly referred to the French publication.

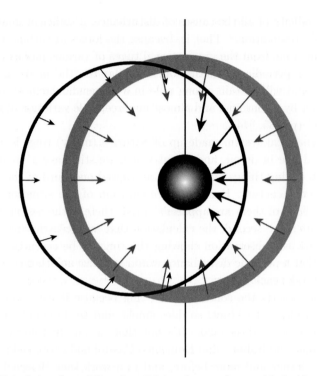

Fig. 3.1   Calculating the stability of a solid ring about a planet, with the idea of applying it to Saturn. We are looking down from one pole of the planet, with the ring (which we assume to be solid) beginning by being exactly centered (thick line). Something pushes it slightly to the left. Now there is more of the ring to the left than to the right of the planet (the vertical line shows the divide), so perhaps there is more gravitational force pulling on it to the right. But the parts of the ring to the right are closer to the planet than those on the left, so maybe the pull of gravity on them is greater, making the ring fall more toward the planet.

centered it is in equilibrium, since the planet pulls equally on all segments in all directions. If it is pushed slightly, so that part is now closer to the planet and part farther away, in what direction is the resulting gravitational force? More of the ring is now on the farther-away part compared to the closer-in part, so just from looking at the amount of ring it might seem that there is more force pulling it back; but the farther-away part feels less gravity, since it *is* farther away, and the closer-in part more, so perhaps that wins out and it keeps moving off-center (see Fig. 3.1).

Going through the mathematics, which Laplace did and Herschel explained, one finds that distance counts for more than amount of ring material, and the ring continues to go offcenter. Thus a ring around a planet is unstable.[8] *Without a mathematical analysis or explanation*, Laplace goes on to say that a hoop which is not uniform, in which the mass is not evenly distributed around the circle, is stable; this is repeated by Herschel. Laplace concludes from this analysis that Saturn's ring is not solid, but fluid. Herschel concludes that the solid ring must be eccentric (the assumption, which is probably a good one, that an uneven hoop would be eccentric is made tacitly). Bowditch gives some amplification of mathematics of the non-uniform hoop, but still fails to give a full stability analysis.

There are two major problems with this calculation, one of omission and one of basic logic. Either reduces it, for our present purposes, to nonsense.

First, there is no mention at all of the possible motion of the ring. If we consider only forces, *all* gravitational situations are unstable: any satellite pushed a little nearer to its planet will feel a stronger force pulling it even closer. That does not mean that orbital motion is unstable. (In fact, both Laplace and Herschel held the Solar System, a collection of orbits, to be stable. We'll go into that in detail after we're done with Saturn.)

Perhaps a better illustration of the effect of motion on stability is a gyroscope, or a toy top. If either is stopped, it falls over quickly. But if they spin they remain upright, balanced on a point. Now, it is not necessarily true that spinning up the ring in this situation would stabilize it (in fact it doesn't), but neither Laplace, Bowditch nor Herschel even considers the problem; even though at least Herschel believes a rotating ring is necessary for other reasons (p. 284, §443).

More difficult to explain is the lapse in logic. Because a centered, uniform ring is unstable, it is assumed (on no other apparent grounds) that a nonuniform, eccentric ring is stable. It would do as well to believe that, because a young lion is a carnivore, an old wolf must be a vegetarian.[9] All in all, this is a mystifying performance by three very capable mathematicians.

Now we come to the observations showing that Saturn's rings are in fact not quite centered on the planet.

The beginning of this story is probably to be found in the *Philosophical Magazine* for July, 1828. A letter from Professor Harding of

---

[8] If, instead of a ring, there is a hollow shell around a planet, the effects of quantity of matter and of proximity exactly cancel out: the shell is neutrally stable. This was pointed out by Herschel (pp. 352-3, §556n), and makes many calculations of the gravity exerted by and on shells and spheres much easier.

[9] It appears that Laplace and Bowditch may have been misled by thinking about the motion of the center of gravity of the ring, which changes position if the ring is nonuniform. And Herschel at least was thinking of a nonuniform ring as approximating a normal satellite. Lapses in logic remain, however, as well as the problem of using an assumption in place of analysis.

Fig. 3.2   An engraving of Saturn, from Sir John Herschel's *Treatise on Astronomy* of
1833. Struve's measurement of the rings being offcenter compared to the planet amounts
to less than four-tenths of a millimeter on this illustration as printed.

Göttingen (Harding, 1828) related how Heinrich Swabe had gained the
impression that the eastern side of the rings of Saturn were farther from the
planet than the western side, and the two had agreed over several months
(December 1827 to May 1828) that it appeared to be so. Harding thought
it was an optical illusion, but could not explain it, and asked astronomers
with better telescopes and more precise instruments to look into the matter. The response of James South to this request (South, 1828) appears in
the same issue. South and several other observers, including Sir John Herschel, observed the planet over a few days in April and May using South's
"five-foot" refractor (the lens of which would have been about five inches
in diameter). Ten micrometer measurements each by South and Herschel
showed no significant difference in the distance between the planet and the
rings, the west side and east side figures agreeing to within a tenth of an arc
second (indeed, Herschel's measurements show the *west* size to be larger,
though by an insignificant amount). However, Herschel (as well as six of
the seven other observers) remained convinced that the eastern side of the
rings looked farther from the planet.

   Not long afterward F. G. W. Struve published his conclusions. In 1826
he had carried out careful measurements of the rings with the Dorpat

refractor (whose lens was slightly over nine inches in diameter and thus significantly more powerful than South's telescope) but had not noted any eccentricity (Struve, 1826). Returning to the question (Struve, 1828), he found that the average of fifteen observations on six nights gave the eastern side of the ring a distance from the planet 0.215" greater than that of the western. He did not publish all his obervations, but reported an uncertainty for a single observation of 0.095," giving an uncertainty for the average of fifteen observations of 0.024" (since, statistically, the average of many observations is more exactly determined than just one measurement). A different approach to the numbers Struve gives yields a somewhat more pessimistic uncertainty of slightly more than 0.161" for one observation and slightly more than 0.042" for the average of fifteen. In either case, his measurement is not much bigger than his uncertainty, while the trick of averaging many measurements places the observation on firm statistical ground. Indeed, Struve declared, "there can be no doubt of the reality of the difference" between the gap to the east and the gap to the west of the planet.

It would take us too far afield to attempt a (probably inconclusive) inquiry into how Struve came up with this erroneous result. From the standpoint of the consumer of data, the preeminent fine-detail observer of the early nineteenth century, using the best instrument of the time, had found a small but significant effect. On the other hand, Struve reported that the average gap between the planet and rings differed by 0.128" between his 1826 and 1828 measurements, indicating a systematic error somewhere; and Short and Herschel had not detected any eccentricity down to 0.1."

Faced with this situation a conservative scientist would consider eccentricity a possibility, but would require further (and, if possible, better) proof before accepting it. Indeed, in his comments on the *Mécanique Céleste* Bowditch considers the eccentricity "probably an optical illusion" (Laplace (1799), p. 492, §44n; recall Bowditch's translation appeared in 1829, after Struve's measurements). Bowditch was a cabable mathematician, not an astronomer, but was also a skilled navigator who had some expertise in making difficult observations, so his opinion has some weight. It is possible (I am speculating) that Herschel had in mind his doctrine of progress by the refinement of errors (Section 3.3), in which a new discovery would initially appear as a discrepancy about the size of instrumental uncertainty.

The last point we need to look at concerning Saturn is Herschel's assertion that a periodic disturbance could stabilize the rings, in the way that a juggler might balance a pole on his finger. The trouble with this idea is that to stabilize a hoop by an *exact* match of disturbing forces to the mode of instability is even harder than managing to center a hoop exactly in the first place; it requires an intelligent and alert *someone* to balance Saturn's ring on a metaphorical finger. It seems very unlikely that things would just happen to work out right, and without some additional analysis on Herschel's part we are justified in doubting it.

It is possible (again, this is only my speculation) that Herschel was thinking of some of the results of planetary perturbations. He spends an entire chapter explaining how the small gravitational influences between the planets change each others' orbits, finding many examples of periodic variations. Indeed, he italicizes the "principle of forced oscillations," in which a periodic change in one part of a system is necessarily imparted to the rest of it (p. 334, §526). However, this is not the same as a demonstration that periodicity alone can conjure stability from instability.

In summary, then, Herschel's exposition on the stability of Saturn's rings depends on a mathematical analysis which is at best inadequate, either not performed at all or irrelevant to the question; supported by an observation that agreed with his visual impression and the erroneous bit of theory, but did not agree with results from other observers. Together with this we have an assumption of solidity that is questionable but that Herschel left unquestioned. I think it's entirely fair to say that he should have done better in each part. The very least we could expect is a caution that the results are not well-established, pending further observation and analysis.

In the absence of such a caution, could a reader have had doubts? I think one of the level of mathematical skill and physical insight Herschel assumed for his readers *could* have noted the defects in the stability argument. But since they were largely matters of omission, a reader might have concluded that Herschel simply not included details of a tedious or difficult nature. There is the clue that Herschel considers that the motion of an eccentric ring occurs "probably under laws of much complexity," that is, the laws are as yet unknown, and so the real behavior hasn't been worked out yet. I think the analogy of balancing a pole on one's finger should have raised doubts; either the analogy is not a very good one, or Herschel is tacity postulating some active agency to counter instability. So there are clues to indicate that the stability arguments are flawed, though they require some mathematical or physical insight and an inclination to question authority.

Whether an observation "of extreme delicacy" should be trusted (Herschel did not give any of Struve's numerical results) is a difficult question, one in which the reader should have been justified in leaving to Herschel.

Finally we have the unexamined assumption that Saturn's rings are solid hoops. The layman is particularly unprotected against this kind of problem. If the experts working in the field haven't thought of a certain idea (in this case, that the rings might be fluid or gaseous or some other kind of structure) or simply don't mention it after spending much time and cleverness on it, it's unlikely that the casual reader will come up with it. If the reader does think of something not mentioned in the text, then most likely it's already been thought of and discarded for a good reason. There is very little the reader can do to guard against this problem. The writer can try to be aware of it, but examining one's unexamined assumptions can take an unusual level of self-awareness.

The last bit of gravitational dynamics we will look at concerns a very big question, that of the stability of the whole Solar System. To this end Herschel gives two results of the French mathematician Lagrange (as presented by Laplace). Although he acknowledges that "it is impossible here to give any idea" of the method used to derive them, he can state them in words as well as mathematical language. The first of the results is that, multiplying the mass of each planet by the square root of its major axis, then by the square of the tangent of its inclination; then adding up all the numbers you get; the sum will be constant, unchanging in time. For those who see this better in formulas:

$$\sum_p m_p \sqrt{a_p} \tan^2 i_p = \text{const.} \tag{3.1}$$

Since all planets are close to a single plane, the inclinations are not large and the sum is small now, and so it must remain. The result thus guarantees "the stability of the orbits of the greater planets" (p. 328, §515) for all time (p. 315, §491). As we've seen, in his book Herschel has gone to great lengths to give a detailed picture of the mutual perturbations of the planets, showing how the various mathematical quantities arise and how it becomes plausible (without actually doing the mathematics) that excursions in inclination will be limited.

The second result of Lagrange sounds similar to the first: take the mass of each planet and multiply by the square root of the major axis, then by the square of the eccentricity of its orbit; adding up all the numbers you will get a sum that remains constant in time. In a formula:

$$\sum_p m_p \sqrt{a_p} e_p^2 = \text{const.} \tag{3.2}$$

Since the eccentricities of the planets' orbits are all small (and by definition less than one, so when they're squared they get even smaller), the sum will be tiny. Again Herschel spends a great deal of time and effort trying to give a picture of how all this comes about, without actually trotting out the mathematics.

It may take a bit of pondering to see how these equations work. For each, you have a quantity for each planet, and a sum of these quantities, the total being constant. That means if one planet acts to increase its quantity one (or more) of the others must decrease. If one planet's orbit gets more inclined, another's must get closer to the average plane.

Similarly, if one planet increases its eccentricity, another's orbit must get closer to a circle. Since the total is small, there is an upper limit to how inclined or eccentric the planets in the Solar system can get. Since it takes a significant eccentricity before one planet's orbit starts to intersect another, allowing close approaches, ejections, collisions and other interesting things, the formulae act to keep the planets separate. (This is the tacit definition of "stable" that is being used in this case. Others are possible.) And since all the terms in each sum are positive, you can't have large pluses and minuses cancelling out.

Consider if the sum in either case happened to be exactly zero. Then all the inclinations or eccentricities would have to be zero also; and they would all stay zero.

Note that in both of Lagrange's results no process is set out by which any exchange of eccentricity or inclination can take place. This is a general limit, and indeed it might happen that no planets ever change their orbits even to the extent allowed by the formula. Herschel's lengthy expositions of mutual perturbations indicate that this is indeed the case.

So the Solar System is stable. These results are (justifiably) looked on as a triumph of analytical dynamics. (They also had important philosophical and theological implications, at least for some people, a matter we won't go into here.) Unfortunately, they are not the firm conclusions that they appear to be. In fact in a recent result (Laskar & Gastineau (2009), with a less technical explanation in Laughlin (2009)) a computer calculation has given a 1% chance that Mercury will be driven quite out of its orbit over the next several billion years, colliding either with the Sun or with Venus or otherwise coming to a spectacular end. How can this result be reconciled with the work of Lagrange?

First, we can look within the formula itself. Note that each term involves the mass of the planet and the (square root of) the size of its orbit. That means Jupiter's term is larger than any other, and any variation in its eccentricity or inclination shows up to a greater degree in the sum; so its orbit is most tightly constrained. A lighter planet or one with a smaller orbit can vary much more before coming up against the limit. As Herschel notes

> There is nothing in this relation, however, taken *per se*, to secure the smaller planets — Mercury, Mars, Juno, Ceres, etc. — from a catastrophe, could they accumulate on themselves, or any one of them, the whole amount of the *excentricity fund*. But that can never be; Jupiter and Saturn will always retain the lion's share of it. A similar remark applies to the *inclination fund*. (Herschel's italics, p. 369, §577n)

A Solar System that allows a catastrophe to Mercury or Mars is arguably something less than stable. It certainly would not appear so to anyone on those planets, however unaffected the others might be. Putting in numbers, Bowditch (Laplace (1799), Book II, Chapter vii, §57, note 762, p. 605) finds that if Mercury, Venus, Earth and Mars were all given an eccentricity of one, that is, thrown out of the Solar System altogether, the eccentricity of Jupiter would be reduced by one-eighth of its present value.[10]

Herschel's remark above — "that can never be" — appears to be founded on his analysis of the mutual perturbations of the planets, as described in his chapter and founded on the mathematics set out by Laplace. Lagrange's results are also derived in the *Mécanique Céleste*, so it is worth while to see where they come from.

As I explained in the last chapter, there are three ways to deal with a differential equation of the sort one encounters in celestial mechanics. Lagrange, Laplace and Herschel used the third method, that of perturbations. That is, in a physical picture, they set out an approximate path as an initial guess at the solution, and worked out how much the change in momentum failed to match the gravity force along this path. Then, from this mismatch, they calculated an adjusted path so that the change in momentum matched the gravity force *along the initial path*. The assumption is that the mismatch between gravity and change in momentum is smaller along the adjusted path.

Mathematically, they changed an intractable expression into an infinite sum of terms, and then assumed that the first terms amounted to a very close approximation of the total. That is, they assumed the series actually

---

[10]The possibility of a change in the semi-major axis, that is the size of a planet's orbit, is not addressed directly by any of our authors. It is certainly permitted by the formulae. In practice, though, it is generally easier for perturbations to increase an eccentricity than a semi-major axis, and disaster will usually come from this direction.

added up to a finite sum, and that the terms they ignored were tiny by comparison.

In each case, the assumption is not always true. Indeed, Bowditch pointed out that some of the terms Lagrange ignored are in fact about as big as those he included. Since the terms involve eccentricity or inclination, one can say (though it's not quite fair) that Lagrange's results show that eccentricity and inclination remain small, if they remain small.

So the stability of the Solar System was not proved after all. Neither, so far, has its instability. It turns out that the answers one gets for long-term solutions of planetary motion depend very critically on *exactly* what one uses as a starting point. This sensitivity to initial conditions is a hallmark of *dynamical chaos*, a subject that takes us long past any book we'll look at. But it means that Laskar and Gastineau could only give a probability for stability (though, at 99%, it looks likely).

Should Sir John Herschel have had doubts about the stability of the Solar System? Certainly Bowditch noticed, as others had, that Lagrange's results were not as tight as one could wish, and Herschel states explicitly that they allowed for catastrophes to smaller planets. Herschel certainly cannot be faulted for failing to forsee the theory of dynamical chaos a century in advance. It took much work and brilliant insight before the assumption that small changes in the input would give small changes in the output, in the field of celestial mechanics, was questioned and finally abandoned. I think the only fault one might charge him with was a categorical statement, that the stabilty of the (major planets of the) Solar System had been proven, rather than giving a qualified and complicated account of what had actually been done (which he had done with his earlier statement on Newtonian gravity and perturbations).

What about his reader? The sort that Sir John was assuming could certainly insert numbers himself or herself into the equations *that Herschel provided*. I think this is important. He gave not only the interpretation but enough detail to check it, at least in principle. In this way, any criticism of Sir John in this matter must at least be muted.

3.6   Summary and lessons learned

In dwelling upon Herschel's problem sections I have given quite an unbalanced picture of his book as a whole. In fact the inaccuracies and mistaken sections form a very small part of it, and none of them vitiate his major themes, expositions and conclusions. Not only is it an excellent description of the state of astronomy one-third of the way through the nineteenth century, most of it can be taken over whole into the present science. In fact the extended explanation of planetary perturbations *without mathematics* has no equal of which I am aware (and I doubt any current author would think it worth the effort to try to match it).

From the standpoint of mistakes made by a scientific author, I can identify seven general sources in Herschel:

- *Reliance on visual impressions.* Very different things can look exactly the same. This is especially pernicious in a science like early nineteenth-century astronomy when almost all the data you have are a telescope view. Herschel was a skilled and experienced observer, and most of the time was cautious about relying on what things looked like; his evaluation of nebulae is a case in point. In the matter of lunar volcanoes and the eccentricity of Saturn's rings he was not cautious enough.

- *Seduction of the elegant result.* The "proof" that the Earth's orbital eccentricity has no effect on its temperature is a very nice bit of mathematical reasoning, exact in result, but wrong in application.

- *Results of an immature science.* In this category I place Herschel's excursions into geology (Sun-driven volcanoes) and thermodynamics (orbital eccentricity has no effect on temperature). At the time, these had nothing like the power and rigor of astronomy, and any results should have been suspect.

- *Categorical statments.* Herschel of course could not forsee quantum mechanics or General Relativity, and in fact their effects were generally unobservable with his instruments, so his proud but qualified statement of the succes of Newtonian dynamics is exactly true as written. It is when he ventures into the categorical, contending that Solar System is stable (without qualification), that he lays himself open to later revolutions.

- *Unexamined assumptions.* This is by far the hardest to detect and guard against. Is there something you just haven't thought of? It is here that a competing theory or explanation shows its value, by requiring an examination of one's assumptions and by pointing out vagueness in formulation and interpretation.

- *Over-interpreted calculations.* While both of the stability analyses we've looked at were mathematically correct as stated, neither actually proved what Laplace and Herschel said they did. In a complicated situation (and astronomy is only going to get more complicated as we go on) a simplified calcuation must often be made, as a way of indicating the way for more complete work; but they must not be mistaken for the final word.

- *Reinforcement of errors.* This is sort of the converse of the advancement of the science by seeing small effects, individually marginal, but convincing when they appear independently in many places; in Herschel's terms, *errors* in an early form of the science that cannot be eliminated by careful work. In this case, however, we have a marginal result (the off-center rings of Saturn) which seems to be reinforced by a theoretical calculation (which is itself in error). This effect is, I believe, vastly underrated by most scientists and unrecognized among nonscientists.

Now from the other point of view, the one I've started with and intend to continue: what can the lay reader take away from the hindsight analysis of Herschel's book?

- *Pay attention to qualifications and expressions of doubt.* They are not there just to pad out a line of blank verse or to demonstrate excessive humility on the part of the author. And apply them liberally; if Herschel says his picture of water-based erosion is only a demonstration of possibility, remember that if it crops up again. If he is baffled by what powers the Sun and unsure of what sunspots really are, his whole interpretation of Solar observations may need drastic revision.

- *Beware of categorical assertions.* If Herschel had said that all motion under the influence of gravity was now completely understood (and it must have seemed a small step from what he actually said) he would have been wrong. When he did assert the categorical stability of the Solar System he laid himself open to trouble (though he must be given credit for giving enough detail in support of this statement for some of his readers to check it independently).

- *Confused or incomplete demonstrations.* If the author's exposition is not clear or there are important parts that aren't mentioned, it might be that his own conception is confused or incomplete, and indeed possibly erroneous. The conclusion may in fact be quite firm, but the author owes the reader a clear and complete exposition of the reasoning behind it. Otherwise, some doubt is in order.

- *Look at the scientist's list above.* You will not always be in a position to determine the reliability of an observation, the elegance of a bit of mathematics or the fact than an assumption has been made, but if you're aware of why they might say incorrect things you may be more alert to detect the possibility.

Chapter 4

Sir George Biddell Airy, *Popular Astronomy*, 1848

The year is 1848. It is a year of revolutions in Europe, though Britain suffers less upheaval than countries on the Continent. Slavery has been abolished throughout the British Empire. In science, Foucault's pendulum has made the rotation of the Earth directly visible and the gyroscope has become an interesting way of demonstrating some of the stranger aspects of rotating motion. The invention of the telegraph has begun to revolutionize the astronomical art of exact timekeeping, though much remains to be done on a practical level. Two years before, astronomy had produced a major triumph in the prediction and discovery of the planet Neptune.

In the spring of this year the Astronomer Royal, Sir George Biddell Airy, is giving a series of six public lectures on astronomy in the provincial town of Ipswitch. Airy is one of the more scientifically active of the men who have held this prestigious position, working in mathematics and mathematical physics as well as astronomy. (His position of authority is underlined by the fact that he felt compelled to explain publically why it came about that Neptune was not discovered first by an English astronomer.) The lectures are written up as a book some years later and appear in a fifth edition as late as 1869, reprinted in 1881. They change very little in substance over that period, however, and we may assign the written form (Airy, 1848) as well to 1848 for all practical purposes.

4.1 Form and purpose

A lecture, in which one person conveys information to an audience by voice, is a very different form of communication from a written book or paper (this point is not always grasped by many a scientist called on to give a talk). The level of complexity of the language must be lower when the audience cannot reread a long sentence several times or flip back to a previous page to check something. Time to assimilate an idea is fixed and generally short. On the other hand, the tone and pacing of the lecturer's voice can convey subtleties that are difficult to put across in print, and one can use a tabletop demonstration apparatus to show things impossible on a static book page.

When a lecture is printed some of these differences disappear but without a completely fresh start the general character will remain. We can expect that a book of lectures will contain an argument of limited complexity and (due to limitations of time) limited total size. Airy's lectures appear to have been straightforwardly written up, without much change from their original form.

4.1.1   *The audience and the object*

Airy is definite about the people he will be talking to. They are "persons concerned ... in mechanical operations" (p. 2), that is, those who make things with their hands; plus as much of the "working-man" as he can reach. While the former implies a significant level of skill and knowledge and indeed includes those who make instruments for astronomers, theirs is not generally an abstract mathematical skill; so Airy will stay clear of algebra (p. 4), for example. This makes a very marked contrast with Herschel's free use of geometry and trigonometry. Airy's audience is much closer to our conception of the meaning of "popular," though he does demand a certain familiarity with a few basic principles (and the patience to go through a long argument in applying them, p. viii).

As to his object in giving these lectures,

> I have proposed it to myself as a special object, to show what may be comprehended, by persons posssessing common understandings and ordinary education, in the more elevated operations of astronomical science.

More specifically, he intends to show that astronomical results are not "mysteries beyond ordinary comprehension" that must be accepted on "loose personal credit" (p. vii). Indeed, it is "most dangerous" for people to believe things only because they are told, and a great failing in the educational system that it teaches in that manner (pp. 80-1). Instead, Airy intends to explain "... so that you may be able to think and in some measure investigate for yourselves" (p. 235).

So Airy seeks to remove the mystery and the function of authority in popular astronomy (thereby doing away, for instance, with my first irony of Chapter 1). In a sense he is trying to *prove* results, something Herschel explicitly discarded as an aim. The difference between their purposes is subtle but real. Herschel sought to provide an understandable, coherent picture of the subject and its workings; Airy wants to show enough to allow his audience to work things out for themselves. It is a more ambitious aim, especially if it is to be carried out on a lower mathematical level, though (partly due to the limitations of the lecture form) it will be restricted to only a part of astronomy's subject matter. On a practical level, we may expect Airy to provide a single clear and solidly-reasoned explanation where Herschel sought a conjunction of several explanations, analogies and descriptions.

4.1.2 *Two themes*

This restriction of the subject matter forms the first of two themes that I identify in Airy's work. He does this by seeking to answer only two questions: how the distances to astronomical objects (beginning with the Moon) can be measured, and specifically, given in terms of something as familiar as feet or yards; and how the weights of the Moon, Sun and planets can be measured and given in terms of familiar pounds avoirdupois. These questions give a structure and coherence to Airy that are impossible for a general survey of the field, such as Herschel produced. It also means that some things of undeniable importance in astronomy are overlooked or given short notice; the prediction and discovery of Neptune, that triumph of mathematical astronomy, appear only in a footnote (pp. 238-9n).

The second theme is the insistence upon personal observation by his audience. "You can learn more by your own observations" than by any lecture or book (p. 88), a point repeated several times (pp. ix, 8). Any modern teacher would heartily agree: a hands-on activity gives much deeper and longer-lasting learning than any paper exercise (to say nothing of an ephemeral classroom lecture). Moreover, even the "coarsest observation with the unaided eye" will show the truth of what he is trying to get across (p. ix).

But *is* the true picture that obvious? Recall that Herschel made the exact opposite point, that the apparent picture and the real were in striking contrast. It seems very strange that the truth should be clear to Airy's working-men and "mechanicals," when it is anything but obvious to Herschel's mathematically skilled readers.

In part the contradiction is a matter of a difference in the emphasis of our two authors. Airy means that observations, giving the basic picture on which his arguments for the measurement and weighing of astronomical objects depend, can be largely done by anyone without refined instruments. He is particularly insistent, for example, that his audience understand diurnal motion from their own experience, and that no progress can be made until that happens. Herschel points out that the arguments and analyses by which one passes from the apparent to the true picture are subtle and require great thought and attention. The difference between them comes mostly from which part of the whole subject they are looking at more intently.

But that does not account for the difference completely. Later on we will have to deal with Airy's opinion of the obvious again.

4.2   Content and results

The subjects of the lectures break down in this way:

Lecture	Subjects	Pages
I	diurnal motion, refraction, transit circle, observations	37
II	size and shape of the Earth, rotation, revolution around the Sun	42
III	motions of the planets, parallax	37
IV	parallax of the Moon, transits of Venus	44
V	precession, nutation, aberration, stellar distances	38
VI	speed of light, proper motion, gravitation, density of the Earth, perturbations, weights of planets	78

All these subjects were treated by Herschel and in roughly the same proportions, though Airy spends more time on the particulars of observing with a transit circle and on the transits of Venus and their importance. Overall (as expected from the limitations of lectures) he has less space for everything than Herschel has in his corresponding chapters. Considering his two guiding questions, we can see that most of his time will be spent dealing with distances; only one lecture of the six (though it appears to be a double-length one) is devoted to weights. Of planetary perturbations, to which Herschel devoted his longest chapter, Airy describes in detail only one type (p. 231), explaining that, "There is nothing in science which presents the degree of complication that these perturbations of the planets and their satellites present." (p. 232)

Of physical astronomy there is no sign. Of course that could be expected from the two guiding questions as Airy has set them out, but it means that we shall not be able to compare his views with Herschel's on the problems identified in the last chapter. Indeed, even when the opportunity arises to set out possible physical ideas Airy declines. For example, he confines himself to saying, "We cannot tell how the planets have been put in motion," (p. 130) rather than bringing up something like Laplace's nebular hypothesis even as speculation.

That does not mean that everything is old and uninteresting. After finding the distance to the Moon in yards, his initial professed goal (pp. 139ff), he goes on to determine the distances to the planets (by way of the transits of Venus, the "most difficult subject for a public lecture" p. 160) and then to the stars.

Recall that finding the annual parallax for any star had not been done at the time of Herschel's book. By 1848 Airy had several possible results to choose from. He sounds unenthusiastic about the result of under 2" for Alpha Centauri, introducing it with a faint, "it would seem" (p. 195), though this more probably reflects his emphasizing to the audience how small all stellar parallaxes are. (The numbers he gives are twice the angle defined in Chap. 2.) Of the smaller figure of 0.1" given for certain other stars, he finds it "almost impossible to answer for" (p. 196) and goes on to list his doubts. Finally, Bessel's result of 0.6" for 61 Cygni was done with "considerable accuracy" and in his "deliberate opinion" it may be trusted (pp. 197-8). His judgement and qualifying phrases here correspond closely with the distances as now known: the actual angle for Alpha Centauri is seven-tenths the figure given; the methods of the time were almost surely incapable of detecting a tenth-second parallax reliably; and Bessel's result was good.[1]

He goes on to present recent results on the solar motion (how the Sun moves relative to the rest of the stars), with doubts (pp. 216-7). Although the results were in fact reasonably close to what is currently accepted, the methods were indeed imprecise and a much more discordant result would not have been surprising.

---

[1] The irony probably escaped Airy that, although his stated aim was to get his audience away from having to rely on authoritative statements from astronomers, he emphasizes his own judgement as a astronomical authority in presenting these results.

Concerning the second of the questions, weighing the planets, as noted above it receives less space. One could speculate that, by the time Airy had finished with distances, his time had begun to run out. More definitely, it's just a harder subject, and more often he is forced to resort to phrases like, "by a process which I cannot hope fully to explain to you" (p. 222) and qualifying a possibly unsatisfying explanation with the phrase "the best that can be done" (p. 236). He does give masses for planets and satellites, reasonably qualified with proper uncertainties. As for the exciting, cutting-edge subject of finding orbits of binary stars, he confines himself to the observation that Newtonian gravity probably applies to them (p. 265).

While my impression is that the descriptions of gravity and the methods of weighing the planets are occasionally unclear and might have been simply baffling to the audience, working out how successful Airy was in his task is beyond my own purpose here. We can only observe that the results he gave are clearly accompanied by doubt when appropriate, so that he is a reliable source for the astronomy he presents even in the light of what is known more than a century and a half later. It is only on an incidental issue, not directly necessary for his exposition or results, that he makes his mistakes.

4.3   The Astronomer Royal mistaken

There are two reasons for spending time on this off-the-track topic. First, it shows a misuse of hindsight, the sort of thing I'm trying to avoid. Second, it demonstrates some signs by which an inaccurate, or at least inadequately known, result may be detected.

Airy makes one extended foray into scientific history, in the form of his exposition of the Ancient Greek system of planetary motion. He says that the system eventually collapsed because of the complexity of ever-growing numbers of epicycles (p. 95); that Ptolemy had no idea of the size of the Earth (p. 144); and that the system survived, in part, because there was resistance to the idea of "demoting" the Earth from its position at the center of the universe. All of these statements are simply wrong.[2] Now, there is no problem at all with a nineteenth-century Astronomer Royal neglecting to learn the proper details of an exploded theory; the problem lies in his

---

[2]For the number of epicycles and the size of the Earth, see (Hoskin, 1997), especially chapters 2 and 4. For the "demotion" of the Earth, Danielson (2009) is an excellent exposition.

proceeding to pronounce upon it in a state of ignorance. The astronomical authority has stepped out of his field, and his statements on history are suspect.[3]

More damaging, I think, and more indicative of problems with hindsight as exercised by Airy, are his statements about the system in general. To him, "It does appear strange that any reasonable man could entertain" such a theory, which is "complicated and unnatural" (pp. 94-5)[4]; indeed, it is "beyond my conception" (p. 98). By comparison, a heliocentic universe operated on Newtonian principles is "much more probable" and Airy can't understand why the ancients didn't figure it out (pp. 69–70). In fact a heliocentric universe under the physics accepted by the ancient Greeks is impossible; you have to get rid of the physics and the cosmology together. To build a consistent Solar System including the dynamics you need Newton, and Airy himself admits that Newton's First Law (at least) is anything but obvious (p. 218).

Going further into this failure of historical imagination would take us too far from our subject. Here I only want you to note the extreme weakness of Airy's arguments: something is wrong, or mistaken, because he himself doesn't understand it. (It is even a step weaker than the support offered by Herschel for the theory he gave of sunspots: no one had thought of anything better.) Generalizing, if a theory is put forward as "obvious" or "likely," or another is rejected as "strange" or "unlikely," you as the reader are entitled to a great deal of doubt — or at least further explanation. And as we noted above in Airy's disagreement with Herschel over "vulgar observations" and the true and apparent workings of the universe, there are things obvious to the Astronomer Royal that might not be so clear to other astronomers (much less people outside the science). He is perhaps too certain of his own position, too unaware of any need to examine his assumptions and conclusions, and too confident in his own exposition; after explaining the variation in speeds of a planet in an elliptical orbit, he assures us

---

[3] I trust you're savoring the irony here: I, an astronomer who has explicitly denied having historical expertise, am telling you not to trust an astronomer's historical assertions. The least you can take away from this is a distrust of the whole matter, perhaps with a resolve to ask a historian. I have given you some references in support of my own statements; I won't be upset if you consider them suspect and instead investigate the matter on your own.

[4] He does not seem to notice the irony of complaining about the complexity of epicycles, while as we have seen he calls the Newtonian theory of planetary perturbations the most complicated thing in science (p. 232).

can do both. The waves we will be dealing with normally vary in both time and space.

Perhaps the best picture to use to illustrate wave motion is that of water waves on an otherwise still pond. If you drop a rock in the water, ripples spread out from that spot. At any given instant of time the height of water varies with position on the pond; at any given position, the height of water varies as ripples move past it.

A wave (see Fig. 5.1) has a *wavelength*, the distance (say in meters) from one crest to the next, which measures how big it is in space; a *frequency*, mesured in cycles per second, which measures how often a crest passes by a particular point; and a *speed*, which measures how fast a particular crest is moving. If you know two of these numbers you can find the third by simple multiplication or division. For light waves in vacuum the speed is always the same, so it's a matter of convenience (or whim) whether you give the wavelength or the frequency, since the other is easily found if someone wants it.[1] In transparent materials (like lenses and prisms) the speed of light varies with wavelength, however. Water waves (especially the larger ocean waves) travel faster with increasing wavelength. Finally, waves have an *amplitude*, which is how much they rise above and fall below their average value. For a water wave, the amplitude is half the vertical distance from the crest to the trough, and will be measured in something like meters; for a sound wave, the amplitude is half the extreme change in pressure as the wave passes. The greater the amplitude, the more energy has been used to create the wave. This is important: for a wave, you can adjust the energy it carries to any figure you want, just by varying the amplitude; it's an analog, not digital, feature. This will become important when we get to the twentieth century.

Waves also *interfere*. If you set one wave and another to pass through a given point in space, their amplitudes add. To make things simple, suppose we have two trains of waves of the same frequency and amplitude, passing through at right angles to each other. If we set things up so that the crest of one wave hits that point at the same time the crest of the other does, we will get a wave of twice the height. A half-wavelength later the troughs will arrive together, and so on, so we'll get a wave twice the size of either. This is *constructive interference*. If, on the other hand, we arrange it so

---

[1] Radio waves are most often given in frequency, so that the numbers on your radio dial indicate so many thousands or million cycles per second, while visible light is more often designated by wavelength, so many nanometers. Organic chemists label their infrared spectra by frequency, but in units of one-over-centimeters, a practice that makes them almost as strange as astronomers.

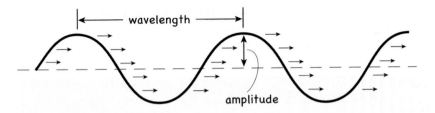

Fig. 5.1 Nomenclature for a wave. It may help to think of this as a cross-section through the surface of a pond, with the water level shown as a set of ripples passes from left to right. Once you have this picture down, then think of how one of these would *interfere* with another passing through a given place at right angles.

that the crest of one wave always arrives at the same time as the trough of the other, we get nothing at all! This is *destructive interference*. The general situation has a pattern of constructive and destructive interference, sometimes very confused-looking, like the surface of the ocean subject to different winds over time.

If you are among a set of waves and moving in a direction either toward where they're going or away, you will measure a different frequency (alternatively, wavelength) from someone who is standing still. Go back to our pond, and have someone make a steady set of waves (maybe by pulling regularly on a half-submerged branch). If you stay still, you'll find a certain number of crests pass you in a minute. If you head toward the source, you'll overtake more crests in a given time, so you'll measure a higher frequency; if you head away, you'll measure a lower frequency. A similar thing happens if whoever is making the waves is moving. This is the *Doppler shift* and it is extremely useful.[2] When astronomers measure Doppler they normally talk about "redshift," which is a change to *lower* frequency and thus longer wavelength (toward the red end of the visible spectrum), or sometimes "blueshift" (depending on which way the measurement went).

One almost metaphysical aspect which becomes very important later on: a wave is not located at a single point. It has some extent in space; it exists over an area or a volume. In fact the simplest mathematical treatment of a wave has it existing *everywhere,* and the more it is confined the more complicated the situation becomes. This is contrasted with the particle

---

[2]The mathematical details relating the amount of Doppler shift to the speeds of the emitter and receiver of the waves are different for different kinds of wave. For sound waves in the atmosphere they're not the same as for light waves in space. But whenever the emitter and receiver are getting closer to each other, the frequency is higher than when they're stationary, and conversely.

description, in which one of the basic things each body has is a well-defined position. The more extended a situation you apply the particle description to, the more complicated the mathematics becomes.

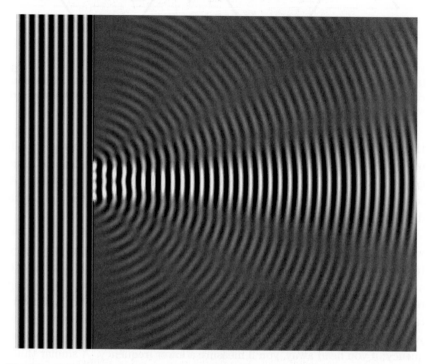

Fig. 5.2   A wave coming from the left encounters a barrier with a hole in it. Think of the bright lines and arcs as the crests of the wave, dark areas as troughs and medium-gray as middle heights. Some of the wave goes straight on to the right, almost like a particle, but there are waves spreading out far to the top and bottom. In this case the hole is four wavelengths wide. A wider hole would create a pattern that looks more like straight-line motion; a narrower hole means the waves spread out more on the right-hand side. A hole much smaller than a vavelength gives a set of almost perfect semicircles on the right. This is a mathematical simulation, courtesy of Richard F. Lyon.

Waves behave very differently from particles in some ways, most obviously when they come upon a barrier. This is illutrated in Fig. 5.2. There is a vertical barrier in the figure, with one hole in it. A particle would either pass through the hole (if properly aimed) or not; in the first case, the barrier would have no effect on its motion. When a wave encounters the barrier part is reflected, but part goes through the hole, spreading out on the far side.

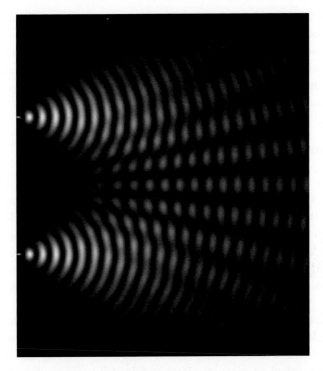

Fig. 5.3   A wave coming from the left encounters a barrier with two holes. Part of the wave passes through each of the holes, and on the other side the different parts of the wave create a complicated pattern. Where crest meets crest, the result is a higher wave; where crest meets trough, the result is nothing at all. In this figure the crests have been highlighted by a horizontal light. Note that there are areas to the right of the barrier that feel the wave in Fig. 5.2, but get no wave action at all here. Figure ©2004-2010 Stephan and Misty Granade, reprinted by permission.

Next consider Fig. 5.3. Now there are two holes in the barrier (the wave to the left of the barrier is not shown this time). A particle would go through one of the holes (or neither); if its location and momentum are known, one could calculate which one it will go through. Or, if the location and momentum are measured afterward, one could calculate which it had gone through. A wave, however, goes through *both* holes. And now look at the complicated pattern of waves on the other side. This is interference, in which crests and troughs of the wave coming through one hole add to crests and troughs of the wave coming through the other. There are places of higher crests and deeper troughs to the right of the barrier than ever appeared on the left (constructive intereference); and there are places where crest has met trough, resulting in no wave action at all (destructive intereference). There are even locations on the right-hand side of Fig. 5.2 that are seeing no waves at all in Fig. 5.3. If you were to set this sort of thing up with light waves, you could position yourself to see a light coming through one hole; then, by turning on another light (opening up the other hole), you could get darkness.

5.2   Electromagnetism

So far the only force we have had to deal with in astronomy is gravity. Others have been known, of course, and Newtonian physics has a place for any sort of force. Two of these forces, known at the middle of the nineteenth century, are electricity and magnetism. To feel either of these you must have a body that is *electrically charged*, in the same way that to have a body feel gravity it must have mass[3]; and to produce either of these forces you need to have charges, as to produce gravity you need to have mass.

But charge comes in two kinds, positive and negative, while mass (as far as we know, which is pretty far) only comes in one. This means you can have a chunk of something that is electrically neutral, and so feels no (overall) electrical or magnetic force, while still having a bit of mass. It also means that the forces are more complicated.

Electrical force is the simplest. A stationary, isolated charge produces a field of force-arrows around it that look just like those of gravity. Look again at Fig. 2.11: the electrical force produced by a positive charge looks just like this — *if* the body feeling the charge is negative. If the body feeling the charge is positive, all the arrows would be reversed. Like charges repel, opposites attract. In fact, much of the mathematics of electrical force is identical with that of gravity, something that saves a great deal of work in solving some equations.

Magnetic force is harder to picture. It depends not only on the charge of the body feeling the force, but also on its momentum, and acts perpendicular to momentum. So while we can draw magnetic field arrows around a source, you can't use them as a direct measure of how to change a body's momentum. What happens most often, if the magnetic field is strong enough, is that the charged particles spiral *around* a magnetic field line (which is what you get when you connect up magnetic field arrows in the proper way).

---

[3]This changes slightly with General Relativity, in the twentieth century.

All this means that a *plasma*, a fluid made up of charged particles or some other sort of charged stuff, can be moved or held up or directed around in ways that gravity alone could not do. In particular, a plasma in the magnetic field of the Earth (which was well known by this period) or the Sun could "levitate" rather than requiring the gas pressure of an atmosphere to counteract gravity (see Fig. 5.4). However, no electrically charged fluids on a large scale were known to the nineteenth century.

Some connections between electricity and magnetism were known almost from the beginning, though mostly they seemed to be quite separate effects. But a few years past mid-century James Clerk Maxwell (we will

Fɪɢ. 69.—Specimens of solar protuberances, as drawn by Secchi. The bright base in each figure represents the chromosphere from which the red flames rise.

Fig. 5.4 Prominences on the Sun, drawn by a capable observer and printed in Newcomb (1878). One popular nineteenth-century interpretation saw these objects as fountains of glowing material flowing through a transparent medium of unknown composition. We now know them to be ionized gas, following magnetic field lines.

meet him analyzing Saturn's rings) found a mathematical way to put them together, so that in a very basic way they were just two ways of looking at the same thing. The deep details of this reach well into the next century, into Relativity and Quantum Field Theory, and there is no need for us to get into them. For our purposes we need only notice that Maxwell's equations mean that a changing electric field produces a magnetic field, and a changing magnetic field produces an electric field, so if you managed to produce a field changing in the right way you could set up an effect that goes on forever. By wiggling a charge properly you could send a series of electric force-arrows and magnetic force-arrows out into space. In fact, you could set up an *electromagnetic* wave. And this is just what light is, and radio waves, and X-rays, and so forth.

This means that light, and radio and all, are waves, so they will experience interference and all other wavelike phenomena. Recall, also, that in an electromagnetic wave the things that are waving are force-arrows, so such a wave can exert force, as well as carrying energy like other waves. This is normally expressed as "radiation pressure" and in our everyday experience is very, very small. As Eddington points out later (Eddington (1927), p. 26), a beam of light strong enough to give a man a perceptible push would vaporize him first. But the intensity of light inside stars can make radiation pressure important (though this is a twentieth-century concern), and light does carry energy.

5.3   Thermodynamics

Thermodynamics has a forbidding reputation as a difficult subject. In part this is well deserved. One must be very clear about concepts and procedures and the danger of winding up with nonsense is real. From a student's point of view, all too often one is told two or three things that are painfully obvious, leading suddenly to a conclusion which doesn't seem even plausible. I think, though, if we leave out the partial derivatives and intricate calculations, we can keep things mostly clear.

We will start out with the concept of *energy*. I have already used the word without being very definite about what it was. Think of it as an attribute, a number, that an object has, that can be increased or decreased in a definite way. Such an abstract idea is not very enlightening, so we normally think in terms of examples. A stone has *gravitational potential energy* when I lift it off the ground; it has *kinetic* energy when I throw it and it is moving. If I just let it drop, some of the gravitational potential energy is transformed into kinetic energy. And — this is important — the *rate* at which one type of energy is converted into another is fixed. Dropping through one foot of vertical distance at the Earth's surface, the stone gains a specific amount of kinetic energy. It cannot gain more; it can gain less, if some of the potential energy gets converted into another kind; but all must be accounted for at the end. Energy does not get created out of nothing or disappear into nothing, just transforms into different kinds. Thermodynamics is often no more than a (not very sophisticated) type of accounting. It still produces some very powerful results.

Energy is *relative*. To calculate how much gravitational potential energy a rock gains or loses, you have to specify how high it starts from and how high it ends up. To calculate how much *chemical potential* energy a tank of gasoline has, you have to specify what chemicals result when you burn it. Gasoline (plus the right amount of air) has a certain amount of potential energy with respect to carbon dioxide and water (which are the products of complete combustion). This energy produces just so much mechanical energy, raising the gas pressure in a car's cylinders by a calculable amount, in turn giving the car an equivalent amount of kinetic energy as it drives down the road[4] (taking into account losses of energy due to friction and heat, which I'll talk about in a little while). To take an astronomical example, nineteenth-century astronomers could figure out how much chemical potential energy there would be in a Sun made up entirely of coal, if it burned completely to carbon dioxide; and could measure the rate at which the Sun was putting out energy in sunlight; and so could figure out how long a Sun's-worth of coal would last. It didn't turn out to be long enough.

An important form of energy (indeed, the kind that started the science of thermodynamics and which has received more attention than any other) is heat. This is the energy an object has by virtue of having a temperature. A low temperature means less heat energy than a high one. To raise an object's temperature, we must add heat energy; to lower it, we must extract heat energy. How much heat energy corresponds to a given change in temperature, the *heat capacity* of the object, depends on what it's made

---

[4]The relative character of energy is why scientists are skeptical about claims to run one's automobile on water. What is the water converted into? It takes energy to break water into its components, hydrogen and oxygen, and it takes energy to go the other way and add more oxygen, making hydrogen peroxide, so there's less than nothing to push the car down the road. Leaving the water as water gives no energy at all. One could in principle run a thermonuclear reaction, converting the hydrogen to helium, and get a great amount of energy, but that's not likely to take place inside an automobile engine.

of, how much there is and what its current temperature is. For a given situation, the heat capacity is fixed. It always takes the same amount of heat to raise a pint of water's temperature from 10°C to 12°C. It takes twice as much to raise the temperature of two pints of water from 10°C to 12°C. It will take less to make the same temperature change for a pint of ethyl alcohol, which has a smaller heat capacity.

Adding or removing heat can also lead to a change in the substance. If we add heat to water at 99°C at first it warms up; then the temperature stays at 100°C, even as we add heat, while the water turns into steam.[5] The heat energy goes into the *latent heat of vaporization*, another form of energy, and does it at a fixed rate (so much energy, so much water goes to steam). To make steam into water you have to extract this energy. If you have nowhere for the energy to go, you're stuck with steam.

When I drop a rock onto the ground, its gravitational potential energy has been converted to kinetic energy, and then — it stops. What has happened to the energy? Part of it goes to dent the ground, but part goes to raise the temperture of both rock and ground. It doesn't rise very much; the "mechanical equivalent of heat," the conversion factor between heat and other forms of energy, is small and difficult to measure. It was one of the early successes of the science of thermodynamics to do so.

Rocks drop to the ground all the time, raising their temperatures a fraction of a degree. But one never sees a rock spontaneously pop into the air, from a place on the ground that has suddenly cooled by a fraction of a degree. From the point of view of energy conversion, the process would be the same. But not all conversions actually happen. The Second Law of Thermodyanamics is the one that forbids spontaneously hopping rocks. There are several formulations of it, one of which says roughly that heat always flows from hot to cold. That's not quite right: in my refrigerator, heat is flowing from cold to hot all the time, to keep the food cool. But in order to do that, somewhere else *more* heat must flow from hot to cold (or the equivalent sort of thing must happen to another form of energy). Taken as a whole, energy flows downhill, though here and there some might go against the flow. Not all energy is available. That's the Second Law.

Next we come to another of those constructions that are extremely useful even though they don't actually exist. This one is the Ideal Gas. Its two virtues are that it is easy to calculate with, since the relationship

---

[5]These numbers are only good for standard atmospheric pressure, pure water, and so on. I'm leaving out a lot of the caveats to keep things simple.

between pressure, volume and temperature is simple; and that many real gases behave very much like it, especially at temperatures high enough, or contained in volumes large enough, or at pressures low enough. (Real gases tend to condense into liquids at low temperatures or high pressures, and liquids don't behave at all like ideal gas.) It is a useful substance to demonstrate a bit of how thermodynamics works.

The best place to keep an ideal gas, for demonstration purposes, is in an ideal cylinder closed off by a leakless piston. The gas is at some particular temperature, volume and pressure; there is just the right force on the piston to maintain the internal pressure. Now we add heat energy through the sides of the cylinder (which can be made insulating or not, as we choose). There are three situations in which the result is fairly straightforward to calculate. First, if we keep the volume fixed (hold the piston in place), the temperature and pressure will both rise, in a proportion that depends on the details of the gas. Second, if we keep the force on the piston fixed (so the pressure stays the same), the temperature will rise and the volume will increase, again in a proportion that depends on the gas. Third, if we hold the cylinder walls at a fixed temperature, the pressure and volume may change. A fourth situation arises if we keep the walls insulating and allow the force on the piston to increase or decrease (ideally, we do this very, very slowly); then the temperature, pressure and volume will all change. Any real situation will be some mixture of these ideal constant-volume, constant-pressure, constant-temperature and *adiabatic* processes, but in order to calculate things scientists will try to break them up into pieces that are close to the ideals.

5.4 Atoms and molecules

We will leave aside some of the Ancient Greeks, who were fond of speculating but not as good at making speculations into good science. Then the idea that all matter is actually made up of indivisible building-blocks, that even air and water are grainy on a small enough scale, starts to be interesting in the early nineteenth century. It began among chemists; no doubt astronomers had no obvious reason to deal with the very small in their investigations of the very large, and something so incompatible with the infinite divisibility of calculus might not indeed have been welcome. But it became unexpectedly useful in the context of thermodynamics.

What I've laid out of that science so far is, in a sense, a mathematical structure that is quite independent of the molecular-kinetic theory I'm about to go into. But the two parts complement each other so well that they are normally combined into a single entity.

The molecular theory is easiest to illustrate with gases. Then the little bits that make up the gas, the molecules, can be thought of as identical units mostly far apart from each other that move at high speeds, often colliding and rebounding from other molecules. Then pressure is interpreted as the combined force of many, many molecules hitting, say, the face of a piston, and temperature is a measure of the speed of the molecules. If you heat up the gas, the molecules will move faster and hit the piston harder: temperature rises, the kinetic energy of the molecules increases, pressure increases. It not only works conceptually, all the numbers work out too. To deal with things like the latent heat of vaporization one introduces forces between the molecules that are relatively weak and short-ranged, only making a real difference when either the speeds are small (low temperature) or they spend a lot of time near each other (high pressure).

The kinetic-molecular picture of heat means that there is an *absolute zero*: when the molecules have stopped moving altogether, you can't get any colder. This is the lowest temperature you can have.

To have something behave as an ideal gas, you need small particles (they take up a negligible fraction of the space they're put in) that interact only via weak, conservative forces, ones that do not much change their kinetic energies. The latter requirement means that they don't convert any of their kinetic energy into some internal form (vibrating like a struck bell, for instance) and that their paths between collisions are free, more or less straight lines. An abstract formulation like this will become important in the twentieth century.

The molecules are moving randomly, but in a sort of calculable way. They don't all have the same speed. Sometimes the collisions will work out so that one molecule goes much faster than the rest, and another much slower. But the faster one will be more likely to run into a slower one, and slow down, while the slower one will be more likely to be run into, and speed up. If the molecules are left to themselves through enough collisions (which happens very quickly in air at room temperature) they will *relax* into a Maxwellian distribution of speeds, which looks like Fig. 5.5. This is a very powerful result: it doesn't depend on what sort of speeds they started with,

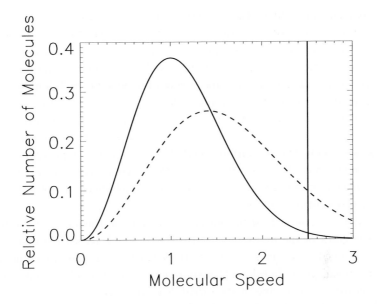

Fig. 5.5 The Maxwellian distribution of speeds in a gas at equilibrium. Speeds are plotted along the horizontal axis, and the probability of a molecule having that speed is the height of the curve. Alternatively, if you have a lot of molecules, the height of the curve is how many molecules have that speed (don't worry about the specific numbers on this graph). Notice that it's very rare to have a slow or stopped molecule; that there is a speed at which the probability is a maximum; and that it's still possible to have a molecule with a very high speed, even though it's unlikely. The dotted curve shows what happens when you double the temperature or halve the mass of the molecules: the maximum goes to a higher speed. More important for some purposes, however, the high-speed end of the distribution gets *much* bigger relative to the lower-temperature curve. Look at the areas under the curves to the right of the vertical line.

or whatever initial conditions you had, eventually this is what you end up with.[6]

Notice that at any temperature the probability of having a molecule completely stopped is zero, but that there's still a chance of having one with a very high speed. This is important. Suppose you have a reaction in which the molecules must hit each other with a certain energy, or above, to start

---

[6]Physicists use the word "relax" in a way that might be unfamiliar and confusing. If you take any sort of system and disturb it, by giving it a kick or by opening a valve or by mixing two gases or whatever, and then leave it alone, it *relaxes* eventually into some kind of final state. It's the process of adjusting to whatever you did, and getting into a condition that will last. Relaxed states are much studied, partly because that's what you end up with eventually, partly because they're generally much easier to calculate.

the rearrangement (call this the activation speed). *Something* will happen even at a low temperature, though it may happen very slowly. Increasing the temperature will increase the number of collisions with enough energy enormously. As a rough rule of thumb, I've heard it quoted that a chemical reaction going at a moderate rate at room temperature will double its rate for a ten-degree increase in temperature.

This may not be what you want. If you need a reaction to proceed at a steady rate for a long time, you want the great majority of your molecules to be well below the activation speed. If you set your temperature so that the maximum of the curve is the activation speed, you'll get rid of almost all your molecules right away. (I've seen respected scientists not notice this, at least in informal talks.)

What one discards with the molecular-kinetic theory is any attempt to keep track of the molecules individually. One deals with statistics, like *average* speeds or (more sophisticated) how many molecules are going faster than a given speed. In taking this approach, we have decideed that only the temperature (which gives a statistical description of the molecular speeds) and a very few other things like the density, pressure and so forth are important. For our purposes, two boxes of gas that have these handful of numbers the same are identical.

There is one more result I want to point out, coming from the connection between the overall temperature and the individual molecular motions. With the kinetic-molecular theory, we can connect the individual kinetic energy of the molecules with the overall heat energy of the mass. If you know how much a molecule weighs, or alternatively how many there are in a container under a given set of conditions, you can figure out how fast (in meters per second) they are moving. Air at room temperature, for example, has oxygen molecules going at something over 400 meters per second (on the average, which is not exactly the same as the peak in the Maxwellian curve). Conversely, if you have some collection of things that interact so as to relax to a Maxwellian distribution, you can assign it a temperature. Sir James Jeans does so in a book we will get to rather later (Jeans, 1929), treating the stars in the Milky Way as a gas. They interact via gravity rather than bouncing off each other, and the relaxation time is very, very long, but they approach the Maxwellian distribution just the same. This calculation gives an extremely high number for the temperature, but there is nothing wrong with doing the calculation.

5.5 Spectra

Although the nature of light was only becoming clear in the second half of the nineteenth century, its usefulness as a way of finding out things at a distance had been established well before that. For this, it was spread out into a *spectrum* of colors by use of a prism or diffraction grating.[7] At one end is red, shading into orange, green, yellow, blue and then at the other end, violet.

It's worth taking a moment to get a clear idea of what the prism or grating is doing. Think of a normal (color) image of an object as a deck of cards lying in one stack on a table, each card representing a tiny range of color. Now take your hand a spread the deck across the table, so that you can make out the numbers or letters on the corners but the cards still mostly overlap. This is what the prism does with colors, except that now they form a continuous band. Each point along the spectrum can be labelled with a *wavelength* or alternately a *frequency*, and in principle between any two wavelengths you can fit another one. (The *resolution* of the spectroscope itself will tell you how close together you can make out features in the spectrum.) It was already clear to Sir William Herschel at the very beginning of the nineteenth century that there is something we cannot see beyond each end of the spectrum. Whatever light is, that which lies beyond the blue end has a shorter wavelength and higher frequency than violet, and is called ultraviolet; beyond the red the wavelength is longer and the frequency lower, in the infrared.

Spectrum analysis started to be useful with Gustav Kirchoff's formulation of three laws. They are

- A solid or liquid, or a gas under high pressure, gives a *continuous* spectrum. This only applies to things giving off light of their own, not reflecting light from another source (as do most things in daily life!). The Sun gives a continuum (with other features we'll get to in a moment), as does an incandescent light bulb. There are different kinds of continuous spectra, but that gets into details we don't need yet. A continuous spectrum is a deck of cards where all cards are present.

---

[7]Historically, prisms were first used; nowadays, for various reasons, diffraction gratings are more popular. The theory behind the way white light is sorted out into colors by each is not important to us at the moment. You can tell a spectrum made by a prism because the blue end is far more spread out than the red. A diffraction grating's spectrum has red, yellow and blue occupying more or less the same length.

- A gas at low pressure gives off radiation only at distinct, separate wavelengths. This is an *emission-line spectrum*. The particular wavelengths are characteristic of the gas's composition; hydrogen always gives the same pattern, for example, which is not the same as nitrogen. An emission-line spectrum is a deck containing only a few specific cards.
- If a continuous spectrum is shone through a low-pressure gas, dark lines appear in the spectrum corresponding exactly with the wavelengths at which the gas' emission lines appear. This is the *absorption-line spectrum*. Later on it was found that the gas must be cooler than the continuum source for this to happen. An absorption-line spectrum is a deck with a few cards missing.

To see some of these things, in perhaps a more convenient way than commandeering a university physics lab, I suggest getting hold of a prism or grating (diffraction gratings of sufficient quality are dirt cheap) and going out some night to look at street lights. Low-pressure sodium lights give a strange, orange-yellow glow which some people find unpleasant; you can trace that to a close pair of emission lines in the orange part of the spectrum. Mercury-vapor or halide lights (much used around sports grounds) have several lines in the blue end. High-pressure sodium lights have something like a continuous spectrum (the gas is no longer at low pressure, and its emission lines are broadened), but with a bit of orange sodium absorption.

So just from the spectrum given off you can tell what material is in a light miles away. You don't have to get close or do any chemical tests. Suddenly a whole new area of investigation is open to astronomers, who (in the nineteenth century) have no hope of ever physically reaching their objects. One of the first subjects was the Sun, in whose spectrum were found hundreds of absorption lines. Many of these were quickly identified with elements known on Earth (though some remained unidentifed for several chapters yet).

I should emphasize that, for the next few chapters, no one knows *why* or *how* an element (also some compounds) produces its pattern of spectral lines. It's not necessary, though, in order to use the phenonenon. And you can even get more information: remember that light, being a wave, will exhibit a Doppler shift if there's a relative motion along the line of sight between the emitter and the observer (that is, if you're getting closer to or farther away from the source) That means you can tell how fast a star is going (along the line of sight) without having to make any precise measurements of its position over decades. (In fact you can't use such measurements

Fig. 5.6 A picture of some city lights at night. The brightest one is a mercury-vapor light (labeled "MV"); most of those visible, in particular four brightish ones in a roughly horizontal row at left (labeled "HPS") are high-pressure sodium. Just above the "IN" label is an incandescent light. (None of this is apparent from picture.) For a view of the spectra of these same lights, see the next figure.

to work out its line-of-sight motion, as you can't use Doppler to work out its sideways motion; the techniques are, in that way, complementary).

I should point out, though, that making a Doppler measurement on light is not easy. If you can measure the position of a spectral line to within, say, 10%, you can measure the speed of the source to within 10% of the speed of light, or to about 3,000 kilometers per second. Stars within the Milky Way move with respect to each other at speeds of about one-tenth to one-hundredth of that, so you need high precision instruments to make the technique work.

I have pointed out that light radiation is one form of energy transfer, that is, that light carries energy. Exactly how it does that and how it delivers energy can be a very complicated and advanced subject, well beyond what was understood for many chapters yet. However, it had already been noticed that there was a greater sensation of warmth, and more effect on a thermometer bulb, slightly beyond the red end of the Solar spectrum, than anywhere else; so infrared radiation is better at carrying heat to Earth-

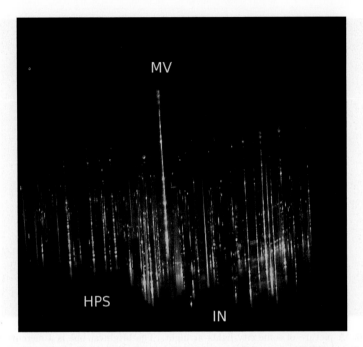

Fig. 5.7   The same view as the previous figure, except that a prism has been placed in front of the camera lens, spreading each light into a spectrum, red at the bottom and blue at the top. The high-pressure sodium lights show emission lines on top of a continuum, like beads on a string, plus one absorption line toward the red end of the spectrum (there is some cool low-pressure sodium in them). The mercury vapor light lacks the sodium absorption feature, but has several bright lines at the blue end. The incandescent light, which starts just to the upper left of "IN," has no emission or absorption lines at all, just a smooth continuum.

bound objects. On the other hand, Sir John Herschel himself had found that the chemical reaction of silver halide, the basis of photography, was not produced by any light in the red and green parts of the spectrum, but that the ultraviolet and blue colors were good at it.[8] The efficiency of these "actinic rays" was very probably on his mind as he formulated some ideas about clouds and the full Moon, which we'll mention in the next chapter.

---

[8]So old photographs, with no sensitivity to red light, show redheads as black-haired and red lipstick also as black. A blue sky, on the other hand, is bright, and indistinguishable from the white clouds in it. It was only decades later that photographic emulsions were made sensitive to green and then red (eventually, in special cases, infrared). On the other hand, the sensors in digital cameras are very good in the red and infrared and comparatively weak in the blue and poor in the ultraviolet. These observations have some relevance to the practice of twentieth-century astronomy.

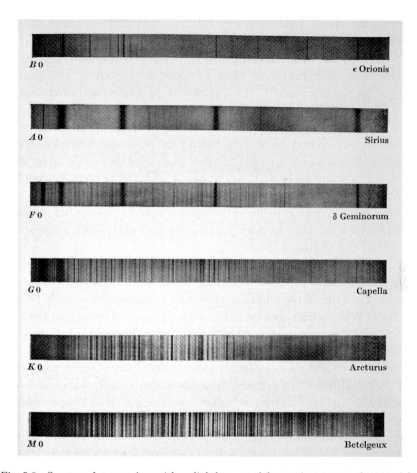

Fig. 5.8 Spectra of stars, taken with a slightly more elaborate instrument than used for the last figures. The image of the star made by a telescope is passed through a thin slit, so that wavelengths close to each other can be separated; it's like using a very narrow deck of cards. Superimposed on the bright continuum are absorption lines of various elements, differing from star to star. The letter-and-number stellar classification, devised in the mid-twentieth century, appears at left, the star name itself at right. These spectra are taken from Jeans (1929); figure ©1929 Cambridge University Press, all rights reserved. Reprinted with the permission of Cambridge University Press.

5.6 On probability and statistics, and keeping track of things

I think these short notes on two unrelated topics will be useful.

I have already cautioned against the use of terms like "unlikely" and "probably" used outside of a strictly mathematical context, that is, as we normally use them. I should clarify this, and add cautions about the use of mathematics in their proximity.

We know what we mean when we say, "it will probably rain tomorrow," (leaving aside the ironic use, as when it's added directly after, "I've planned a picnic for the afternoon"). It's a matter of judgment, and unless one is strict about checking the weather forecast has no numbers behind it. So when a scientist dismisses something as "improbable" or "unlikely" with no numbers, there may be nothing behind the statement beyond personal prejudice. When we set up a simple numerical situation we also know what we mean: it's not likely we'll roll a one in a single go on a die, but we wouldn't be astonished if it happened now and then. Similarly, rolling anything but a one is quite likely, though one wouldn't like to bet the farm on it. And it's not hard to work out that the first probability is one-sixth, the second five-sixths.

It's not so easy to work out the probability of many things, each of which is improbable. I will use my favorite illustration of the poorly-shooting duck hunters. Each one of them has one chance in a thousand of actually hitting the duck. When a duck flies by, it has a good chance of living past the first hunter, and the second. But what if there are a hundred? A thousand? The calculation is not trivial: a thousand hunters at one-thousandth chance apiece does not add up to a certain duck dinner; there is still a possibility that they'll all miss. In fact the duck has about a 37% chance of living through the barrage.[9] My point is that combining a lot of low-probability things has to be done with care; you can neither say, "the hunters are poor shots, so the duck is safe;" nor, "there are a lot of hunters, so the duck is dinner." And astronomy is full of situations which are individually improbable, but where there are many chances.

Statistics happen, more or less, when you dump a lot of probabilities together and try to make sense of the whole. They answer questions like, "I see lots of stars of the tenth magnitude and fainter in this one-degree field of view, but none brighter. Is something going on here?" Depending on the size of the field, how stars are distributed among the various magnitudes *and how many fields you have looked at* the answer could be, "It's quite expected; the chance is 89% you'd see this by now," or "Strange! The probability of this is less than 1%. Let's look into it." I mention this only to point out that answers in this case are both probabilistic (percentages

---

[9]It would take us too far afield for me to tell you how to do this, but if you're intrigued and have some skill at logarithms you can give this a try.

are attached to them: there are no absolute certainties) and normally take some attention and skill to calculate correctly.

A short note on catalogues now may prevent some confusion.

Up until about the middle of the nineteenth century most of the objects astronomers would discuss were bright and easy to refer to: the planet Saturn, for instance, or the star Beta Lyrae. But Struve had already made up lists of hundreds of double stars, many of them too faint for anyone to have given them a name or number, and they were denoted by their catalogue number: Struve 226, for instance. Sir John Herschel and his father had collected lists of hundreds, and then thousands of faint nebulae. Each of them had a Herschel catalogue number, for instance H I 217. Herschel numbers are not now used, since their listing was reorganized into the New General Catalogue and Index Catalogue; you'll see instead NGC 7793 or IC 1613. About a hundred of the brighter nebulae had already been listed by Charles Messier, so that they wouldn't be mistaken for comets, and some had common names. So the Great Nebula in Andromeda is M31 and NGC 224; the Orion Nebula is M42 and NGC 1976; the Ring Nebula is M57 and NGC 6720.

And by this time there were many more observers, many of them making more careful and repeated observations of a few objects, some of them known only through some catalogue number. So from now, catalogue numbers will be found more and more.

To help you out a bit: if you come upon a reference which has a Greek letter, or a number, plus the Latin genitive of a constellation name (Alpha Canum Veniticorum or 55 Andromedae) it's a star. A number following M, NGC or IC will be a star cluster or a nebula (and the latter may be a galaxy or a diffuse cloud of gas).

We needn't worry, yet, about the many catalogues of variable stars, radio sources or things found by X-Ray satellites (or worse).

5.7   More orbits, plus squares and cubes

We've talked about the elliptical orbits of the planets and about some perturbations of them. It's time to discuss some trajectories that are not very elliptical, especially some dealing with three bodies moving around.

We will start with another venerable ideal object (that is, a useful thing to think about but that actually does not exist), an extremely powerful cannon on a mountaintop above the Earth's atmosphere.

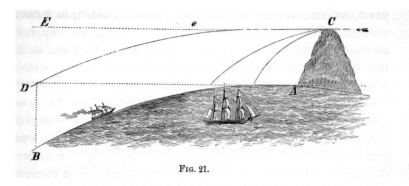

Fig. 5.9   Newcomb's version of Newton's cannon, from Newcomb (1878). The weapon is located on a mountain (which doesn't really exist) above the Earth's atmosphere, so we need not worry about air drag, and can shoot its projectile with any chosen speed. If loaded with a small charge the cannonball goes out some distance and eventually hits the ground (or, here, the sea, worrying the passengers and crew of the sailing ship, on which the gunners appear to have achieved a good straddle); if the ball is sent out at great enough speed, it continues around the Earth.

If we charge the cannon with a small amount of powder and fire it horizontally, it sends a ball out sideways but eventually the Earth drags it down by gravity and it hits the ground. With a larger charge the ball goes farther before striking the ground. With enough speed, however, the fall of the ball exactly matches the curve of the Earth, and it never does hit the ground. The ball has gone into orbit; it takes about seven kilometers per second of muzzle velocity[10] to do this. Seven kilometers per second is the Earth's *orbital speed* (at the Earth's surface; at a higher altitude it is smaller).

If we fire the cannon below the horizontal but just at orbital speed, it will again hit the ground (if the Earth's surface weren't there, but the mass still existed at the center, it would have gone into an elliptical orbit that reached above and below the current radius of the planet). If we fire it so the ball goes straight up, it will get to some high altitude but eventually turn around and fall back again. Think of a non-circular orbit as a continual exchange between kinetic energy (which the cannonball has due to its motion) and potential energy (which the cannonball has because

---

[10]This is a term that artillery people use. According to the purist in scientific terms it should be "muzzle speed" since "velocity" includes direction also (remember the velocity and momentum arrows), but the term is hopelessly entrenched. Note also that seven kilometers per second is beyond the capability of any real cannon made now, much less a black-powder weapon of Newton's time. This is a conceptual cannon.

it is at some altitude above the Earth). It can trade kinetic for potential, slowing down as it gets higher; or vice versa, falling and gaining speed. The exhange rate between kilometers per second and kilometers away from the Earth's center is fixed. So at seven kilometers per second it can exchange all its kinetic energy for potential and still not get away from Earth entirely; that takes a little over eleven km/s.

If we increase the powder change in our cannon and shoot the ball straight upward at 12 kilometers per second, it will still slow down with height, but never to the point of stopping or coming back. It has then reached *escape speed.*

(Now put together the Maxwellian distribution, above, with the idea of the escape speed. The molecules at the top of the atmosphere of a planet will have a certain temperature. If there are many of them above the escape speed, the atmosphere will quickly leak off into space. If the escape speed is well above the great majority of the distribution, the leakage will be slow. Recall also that the speed depends on the molecule's mass: heavier molecules travel more slowly. So the lighter gases, hydrogen especially, will only be found in cold planetary atmospheres, and those of the more massive planets. Small planets near the Sun will only retain very heavy gases, if any at all.)

Recalling the various shapes of orbits, a cannonball that has less than escape speed will be on a closed (elliptical, possibly circular) orbit around Earth's center. If it has less than orbital speed its ellipse will at some place (maybe everywhere) be closer to the Earth's center than the Earth's surface, and it will hit the ground. If it has escape speed it will have a parabolic or hyperbolic orbit, and will leave Earth (as long as we take care the orbit doesn't intersect Earth's surface at the outset) entirely. This works the other way too: if something comes from far away, so that (in a manner of speaking) it has already escaped from Earth, it can approach on an open trajectory and pass very close; but unless it actually hits the Earth it will then head away again and escape.

What happens if we escape the Earth? Well, there's still the Sun to contend with. It is much more massive and its gravity far more powerful. We can escape Earth at 12 kilometers per second, but we'd still be in orbit around the Sun. In fact we are actually in orbit around the Sun all the time. Even the Moon, looked at from a Solar System perspective, is in orbit around the Sun, an orbit slightly modified by the pull of the Earth. As Herschel noted, the gravitational force exerted by the Sun on the Moon is greater than that exerted by the Earth. And yet (this is important!), it is

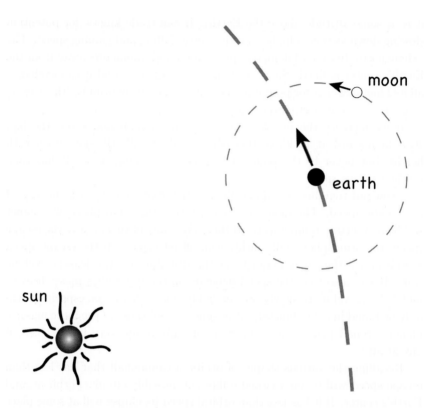

Fig. 5.10   The Moon in orbit around the Earth, with both in orbit around the Sun. You could give an object on the Moon enough speed to send it into orbit around that body, and it would then be gravitationally bound to the Moon, Earth and Sun all at the same time. With a bit more speed it could escape the Moon, but still be in orbit around the Earth and Sun. With even more it can escape the Earth and go into a separate orbit around the Sun, somewhat modified by the gravity of the Moon and Earth.

still accurate to say that the Moon is also in orbit around the Earth. It has not reached its escape speed relative to Earth. The Moon is *gravitationally bound* to the Earth; both are gravitationally bound to the Sun.

What happens to something on an eccentric orbit in the Solar System, say a comet? Looked at from the Sun, it spends most of its time on a close-to-elliptical orbit. But suppose it makes a close passage by one of the main planets. From the planet's point of view, it's coming in from far away, too fast to be bound, and so follows a parabolic or hyperbolic orbit. But the planet itself is in motion relative to the Sun. If the comet passes by on the rear side of the planet (as it goes around in its orbit), the planet's gravity pulls it *faster* along and it gains energy; if it passes on the front side, it loses energy. (It can also pass over one of the poles, slinging it out of the plane of the Solar System.) In any case, its orbit can be so modified as to make it an entirely new ellipse. And if it gains enough energy, it can be flung out of the Solar System entirely. (This means that the term "gravitationally bound" has to be used with care when there are three or more bodies.)

These kinds of "gravity boost" maneuvers are now routine for space probes sent out into the Solar System, and so are familiar to us as they would not have been to any of the authors I'm examining. At the same time, comets were well known and the ideas and (to a certain degree of accuracy) the calculations involved were entirely accessible to nineteenth-century astronomy, there for the taking if the occasion came up.

It may be useful here to explain something about squares and cubes, or more specifically inverse-squares (the mathematical form of the law of gravity and of the intensity of light) and inverse-cubes.

The magnitude of the force of gravity falls off as one over the square of the distance. That means that the farther you go from the source, the weaker the force is, in a specific way. A more massive body will have a

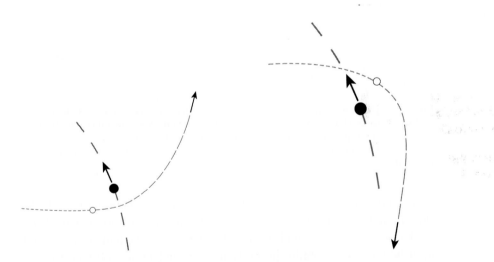

Fig. 5.11 A small object passes close to a planet, with the latter in orbit around the Sun. From the point of view of the planet, this object (a comet, say, or nowadays a space probe) comes in from a far distance and follows a hyperbolic, gravitationally unbound trajectory back to the far distance. From the point of view of the Sun, the object can either (left) gain some orbital speed from the planet, and go off faster than it came in; or (right) lose some orbital speed, and go off slower.

stronger force at a given distance, but its force will also get weaker with distance. The same thing happens to light: whether from a lighthouse or a candle, the brightness falls off with distance in an inverse-square way.

Other forces are not inverse-square. The tidal force of a body, say that of the Moon on the Earth's oceans, falls off as one over the third power of the distance. That means that tides get weak much more quickly than gravity itself. Have a look at Fig. 5.12.

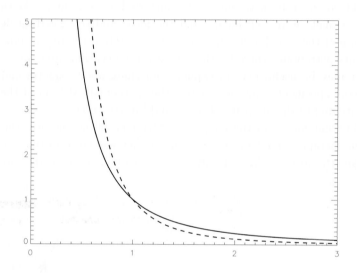

Fig. 5.12   The mathematical behavior of inverse-square and inverse-cube functions, important to things like gravity and tides. The horizontal axis is distance, increasing to the right. The solid curve is an inverse square law, getting weaker with distance. The dotted curve is an inverse cube law, getting weaker with distance (and stronger with proximity) more quickly.

It's fairly clear from the graphs that the inverse-cube curve is higher close in and lower farther out, compared to the inverse-square one. It may not be clear that, if you multiplied either curve by some number (stretching its height in some definite proportion), it would *always* be true that the inverse-cube curve is higher close in (closer than some distance, which one could calculate if necessary) and lower farther out. The inverse-cube curve is always steeper overall, in this sort of general way. That means, among other things, that there is some distance beyond which you can pretty much ignore tidal effects while still worrying about gravity itself.

The same sort of thing happens with squares and cubes: the cubic curve is steeper overall. I find that, while many of my authors will occasionally mention that something happens because one thing is inverse-square and another is inverse-cube (even authors who mention no other mathematics, and leave formulae to footnotes or appendices), none bother to explain it. It's a bit of mathematical intuition that scientists pick up early in their education, and I suppose may not realize that it's not obvious.

The same sort of thing happens with squares and cubes, the cubic curve is steeper overall. I find that, while many of my authors will occasionally mention that something happens because one thing is inverse-square and another is inverse-cube (even collars who mention no other mathematics and leave formulae to footnotes or appendices), none bother to explain it. It's a bit of math/mathematical intuition that scientists pick up early in their education, and I suppose may not realize that it's not obvious.

Chapter 6

Sir John Herschel, *Outlines of Astronomy*, Tenth Edition, 1869

The year is 1869. Travel by train is commonplace, even in the relatively uncivilized countries of the Americas, and iron is routinely made to float (in the form of some warships). Even space has been conquered, as balloons have explored the cold and thin air far above the Earth. There is no doubt that the age of wood and canvas, horses and leather is being replaced by the Age of Iron and Steam. The pace of change in daily life has grown quite dizzying.

In science, changes even more fundamental are occurring. Gustav Kirchoff and Robert Bunsen have shown that one can determine what a substance is without touching it, weighing it, or testing it in any way, just by looking at the light it gives off or absorbs; this will transform observational astronomy. The work of a number of scientists such as Sadi Carnot, James Joule and William Thompson is putting together the obvious and subtle pieces of the new science of thermodynamics, which will transform the interpretation of astronomical observations. And James Clerk Maxwell has just unified the theory of electricity and magnetism, perhaps the greatest feat of nineteenth-century science, though the full implications are far from clear yet. As a sort of warm-up for this brilliant physicist (to use an anachronistic word), a decade ago he investigated the rings of Saturn in thoroughgoing rigor, showing that solid or fluid rings could not be stable, and they must be made up of a cloud of distinct solid or liquid particles.

In astronomy, the size of telescopes has grown quickly; the lenses of the largest refractors are now approaching or surpassing two feet in diameter, while the Earl of Rosse has broken the Herschel family monopoly on enormous reflectors and constructed one with a main mirror diameter of six feet. There is an accompanying increase in quality, harder to quantify but evident to observers. And there is an increase in the sheer number of careful observers with good instruments.

In the midst of all this activity, heading out ever faster in new and unexpected directions, Sir John Herschel has brought out the tenth edition of his *Outlines of Astronomy* (Herschel, 1869) for the general public.

6.1   The new edition

Herschel is now one of the grand old men of British astronomy, and indeed of astronomy everywhere. His contributions to the science have been generally of a different character from those of his illustrious father but arguably as important. He has catalogued the southern sky from the Cape of Good Hope and has presided over the Royal Astronomical Society. He has originated a way to calculate the orbits of binary stars and has done innumerable calculations of his own. He corresponds on a cordial basis with any astronomer of note in the world. He has also done effective work in other sciences, and indeed has been instrumental in making photography practically useful. The new edition of his book can expect to benefit from his decades at the heart of this very active science.

Notwithstanding the different title and the passage of some thirty-six years, perhaps the most important point about the 1869 edition is how much in it is identical with the 1833 edition. Many sections, amounting to roughly half the whole, are word-for-word the same. This should be very surprising. With astronomy changing so rapidly, the fact that most of the very words written before the reign of Victoria should be useful in their fourth decade seems strange indeed.[1] But recall what I emphasized in a previous chapter: most of what Sir John wrote is in fact still true, still accurate, and (especially if one takes to heart his qualifications and caveats) even instructive today. (I think it also has a degree of exactness and felicity of phrasing that cannot always be found in later writing, though I have no intention of arguing its literary merits here).

However, the fact of this being a tenth edition, with the basic parts having been written decades ago and the science having progressed since, means the writing will almost necessarily be uneven. An introduction and

---

[1] In fact, the edition I have at hand was printed in 1902. By that time, of course, the publisher had to add a preface to the effect that it was not a good picture of the *current* state of astronomy. But I'm sure any author would be quite happy to think that a paragraph of his was worth reprinting, without change, more than *seventy years* after it first came out of his quill pen.

general picture of an area written for one state of the science will inevitably fit less well after things have changed; even if all the statements are still literally true, emphasis will have shifted. The author is faced with the choice of leaving things as they stand, if still reasonably accurate; modifying words or sentences here and there as necessary; or rewriting a chapter entirely from scratch. In his 1869 book compared with the 1833 edition Herschel has done each of these things. Some of the old sections are annoted as (for example) "So in the 1833 edition," which appears to mean that things are differently understood now, but for some reason the old wording is still valuable (he does not explicitly say what he means). Some sections have changes made, more or less extensive. And some are completely rewritten.[2] There is also a great deal of entirely new material, which we will look at in some detail below.

6.1.1 *Audience and method*

Recall that Airy had pitched his exposition at a lower level than Herschel's first book, requiring less mathematical understanding and (largely from the limitations of the lecture format) using less complex and intricate arguments. This new approach was at least moderately successful, since the fifth edition of Airy's work is being printed as the tenth of Herschel's comes off the press. But clearly Herschel has not succumbed to the temptation to move in on Airy's public. Indeed, he has included a derivation of an equation in several lines of algebra; quoted several more bits of untranslated French and Latin; and greatly expanded the most intricate section of the book, on planetary perturbations. As he says explicity, comparing the first edition of *Outlines of Astronomy* to the earlier *Treatise,*

> The chief novelty in the volume, as it now stands, will be found in the manner in which the subject of Perturbations is treated ... The chapters devoted to it must ... be considered as addressed to a class of readers in possession of somewhat more mathematical knowledge than those who will find the rest of the work readily and easily accessible; to readers desirous of preparing themselves ... for a campaign in the most difficult ... of all the applications of modern geometry. More especially they may be considered as addressed to students in ... university ... (pp. 11–12, Preface to the First Edition)

---

[2]It is especially difficult to modify earlier writings when the technology of book-printing makes revisions laborious and expensive. Herschel often resorts to adding a miscellany of notes at the end of the volume, rather than interrupting the flow of exposition as established. Added to this is the psychological difficulty of changing or replacing long-established prose, which after years of standing sometimes seems to be the only way something could be said.

Herschel has moved slightly, but significantly, upmarket. Indeed, thinking in terms of current times it might be more appropriate to compare this book with an introductory university textbook than with something written for the general public.

Within this change, his method and plan of treatment remain the same as before (and the sections explaining them are taken over whole from the 1833 book). He is still aware that reasoning on the basis of his explanations, without the mathematics, must be done carefully or one can get the wrong answer (p. 549, §647n). He thus is far less confident than Airy that his expositions will lead people inevitably to the truth. He has been misquoted and misunderstood (p. 715, §798n), rather to his annoyance. It is probably no comfort that other astronomical authorities also have mistakes in print (though in Arago's case, p. 690, §776n, it appears to be from carelessness of oversight rather than any scientific error).

The mention of printed mistakes and other astronomers brings out a significant difference between the 1869 and 1833 editions. In the later book there is much more mention of disagreement and debate on scientific points among the various authorities. Herschel differs with François Arago on the possibility of a faint star being the illuminating source for a planetary nebula (p. 794, §877), with Struve on the form of the Milky Way (p. 709, §793n), and with H. W. M. Olbers on the possibility of some light-absorbing material in space (p. 714, §798, which we will look at in detail below). The reader has a much greater sense of being present at an exciting time in the progress of astronomy, with much going on and much as yet uncertain. It is also much more confusing.

6.1.2 *Content and themes*

To give some definite form to the changes between the editions, I show below the number of pages in each chapter in the 1833 edition, together with the change in the number of pages devoted to the corresponding chapter in the 1869 edition with a correction included for the difference in the size of the pages. Keep in mind this is a rough measure and does not take into account details in the textual arrangement or in things like wording.

Of all the subjects, only one has less space in its dedicated chapter in the later edition, and that comes from the fact that it is a short chapter from which material has been moved to other sections. No subject has actually received a shorter treatment in the 1869 edition.

Chapter	1833 pages	1869 change
Earth	61	+46%
Instruments	43	+23%
Geography	50	+20%
Uranography	27	+7%
The Sun	29	+48%
The Moon	19	+42%
Gravity	11	−27%
Planets/Solar System	45	+24%
Satellites	12	+33%
Comets	12	+225%
Perturbations	60	+145%
Stars	36	+186%
Calendar/Time	8	+200%

Two have about the same amount of space, Gravity and Uranography. Four have a moderate increase, of one-fifth to one-third: Instruments, Geography, the Planets and the Satellites. Three have a significant increase, amounting almost to half again as much material: the Earth, the Sun and the Moon. Finally, four have enormous increases, making them somewhere around three times their former size: Comets, Perturbations, Stars, and the Calendar. In fact the chapter on Perturbations has been split into three chapters, and what went into the Stars chapter in 1833 has also been made into three separate chapters: on Stars (including much on the structure of the Milky Way); Variable and Binary Stars; and Clusters and Nebulae. The work overall has expanded by about two-thirds, most of it coming in Perturbations and Stars (two chapters in 1833, six in 1869).

The much larger treatment of perturbations does not stem, I think, from any corresponding advancement of that aspect of astronomy in the intervening time. Herschel may have been confirmed in his intention to do it by the discovery of Neptune (1846; the First Edition of *Outlines* came out in 1849), a triumph of that branch of the science, but the techniques and applications were already in place. Likewise, his earlier effort at explanation was at hand and only needed to be carried over, as so much of the *Treatise* was. I rather think that he was encouraged to improve and perfect his treatment because it was something no one else had attempted in quite the same way and which had already enjoyed some success. In addition, he is now writing for university students, and feels he must be more careful. A somewhat simplified explanation in the 1833 edition (pp. 348-9, §550), in which the motion of the Sun in response to Saturn and Jupiter is neglected, is now corrected (pp. 624-5, §723n). The earlier version did not actually lead to an erroneous conclusion or even to an inaccurate picture, because its approximation was covered by averaging; but Herschel feels he needs to set things straight (and appears somewhat embarrassed about it).

The lion's share of the expansion, though, consists of observations. There are pages and pages of observations of all objects in the Solar System, comets in particular described in detail; and in stark contrast to the 1833 edition we have extensive descriptions of the distribution of stars within the Milky Way, the two species of star cluster and various nebulae. I think it's worthwhile to consider some possible reasons for this explosion of observations. It is something of a digression from our main question, but helps to provide context for Herschel's book.

First, telescopes had roughly doubled in size. As noted above, the leading refractors had gone from about a foot in diameter to slightly over two feet, and the Earl of Rosse had surpassed the Herschels' family records on giant reflectors, eventually building one with a mirror of six-foot diameter. Much fainter objects became visible as well as fainter details of known objects. Within the limitations of optical quality and especially the steadiness of the atmosphere, the larger instruments could also support higher magnifications, and hence show smaller details. In addition, several of the larger instruments are now to be found in places with better observing conditions than (for example) the Slough of the elder Herschel.

Perhaps more important for the volume of observations, the size of telescopes of the second rank increased. With several refractors of the one-foot class now in operation instead of one or two, there will simply be more significant observations made.

Second, the quality of refractors (still making up the majority of telescopes) improved steadily over this period. It is not as obvious an advance as that of sheer size, but the better design of lenses and especially the production of high-quality glass in larger sizes made later telescopes more effective. (Perhaps one has to spend some eyepiece time in telescopes from the 1830s and the 1850s to appreciate the improvement properly).

Third, the overall number of observers increased significantly. The sky is a very big place, and more people watching means fewer interesting things escape undetected.

Fourth, there were the efforts of Herschel himself. The most tangible consist of the great catalogues of objects (nebulae and double stars) put together by Sir John from his family's observations, covering the whole of the sky, which gave other observers a concise idea of what there was to look at and where to find it. In an era that can call up a map of anywhere at the click of a mouse button, this is worth emphasizing. The sky is an enormous place, and until you have at least a foggy idea of what's out there and where it is, you can spend most of your time wandering around empty space. Once the stars and nebulae were given positions and rudimentary descriptions other observers could follow up with detailed and careful observations.

But Herschel was also a great correspondent, collecting and disseminating information from observers everywhere. So not only was there much more well-directed observing being done with bigger and better instruments, he was in a position to know about it, and collect it for his book.

The expansion of the chapter on the calendar has no obvious astronomical explanation. There is a detailed explanation of how to determine the time between two dates, which perhaps might have been prompted by the need to relate two or more observations to each other, now that there are so many more of them. It is a large proportionate increase, but not great in an absolute sense.

The themes, major and minor, I identified in the 1833 *Treatise* are all still present in the 1869 *Outlines*. Indeed, the idea of intelligent life elsewhere is more prevalent, corresponding to the greater amount of observational material on stars and planets; It is even brought into the discussion of the violently variable star Eta Argûs (now Eta Carinae; pp. 742-3, §830). Another idea present in the early book is the assumption, never made explicitly, that all the planets have solid surfaces (for example, p. 428, §513, where the dark bands of Jupiter are assumed to be the surface). It becomes interesting in the later book since it remains, even applied to the Sun in spite of the fact that it does not result in the proper kind of "trade wind" structure in motions of sunspots (p. 855-6, §387dd). This could perhaps be identified as a failure of imagination: the idea of a body gaseous all through, or with maybe a smooth transition from gas to liquid to solid, simply did not occur to anyone. The lack of anything to hint at such a thing on Earth is an obvious drawback; perhaps also the undeveloped state of thermodynamics made the more extreme states of matter difficult to envision.

With all the new and expanded material in the tenth edition, Herschel's *Outlines* retains an old-fashioned sense of time and patience. A curious reader is assured that it will take only a half-hour's calculation to check a certain result (p. 671, §771), and that a mere thirty or forty years of further observation will surely settle a certain observational question (p. 776, §861).

It is probably safe to say that no current popular book on astronomy could expect a reader to dedicate a comparable time to a calculation (even among those that expect readers to do any calcuation at all), and that current funding bodies are highly unlikely to support an investigation whose result would only be known decades from now.

In summary, Herschel's new edition continues the plan and method adopted thirty-six years before, and indeed includes most of the previous book. The version of 1869 adds much material, mostly in the form of a more detailed description of planetary perturbations and in masses of new observations. We must now look at the Tenth Edition with hindsight.

6.2　Old problems

What has happened to the problems identified in the 1833 *Treatise*?

The mistaken calcuation of insolation, the conclusion that the temperature of the Earth is completely unaffected by the annual changes in its distance from the Sun, starts out as before (pp. 295-6, §368); but this time Herschel realizes that the important point is how much energy is received in a given time: "At about the southern summer solstice then, the whole Earth is receiving *per diem* the greatest amount of heat it can receive ..." and so the southern summer is hotter. He produces some anecdotal evidence for this; perhaps his stay at the Cape of Good Hope suggested the idea. He notes that actual measurements of temperature on Earth are affected by the different responses of land and water to the Sun's radiation (pp. 301-2, §370), though he does not realize just how complex things can get. He even makes a calculation of how much the difference in overall summer temperature between the hemispheres should be (p. 299, §369a). This is wrong in detail, since he is not yet clear on the relationship between heat and temperature, but he qualifies the result with phrases like "there is good reason to believe" and "at least as low," rather than stating flat certainties. A more complete and clear understanding of this kind of thermodynamics was in principle available at this time; Hermann von Helmholtz indeed presented a very well-done exposition for the public in 1862-3 (Helmholtz, 1863). Attention to the more recent work in this field would have helped Herschel but he appears not to have assimilated it, and as we shall see had his doubts about the new theory. In any case we can say he has corrected his earlier mistake and gone some way toward setting out the right answer.

On the matter of the stability of the Solar System there is no change. Although the chapter it appears in has been extensively rewritten Lagrange's results are quoted as before, and the stability of at least the major planets (p. 601, §701n) for all time (pp. 515-6, §604) asserted on that basis. There does not seem to have been any important work on the subject produced in the period bewteen Herschel's books, to draw his attention and to be included in his exposition. The prospect of dynamical chaos did not rear its ugly head until the work of Poincaré at the end of the century, so this is no surprise.

The craters of the Moon are still identified as volcanoes (p. 358), and for the same reason: they look like volcanic craters on Earth, and there is no competing theory. On this subject also much time must pass before any movement can be expected.

The Sun is still seen as the ultimate source of volcanoes on Earth (p. 331, §399), in a section that has not changed significantly. As before, it is reasonable to consider this as only a possible story, covered by Herschel's earlier reservations about water-based weathering as a driving force in geology. In addition, Herschel appends a footnote, "So in the edition of 1833." As I've noted, he does not explain exactly what he means by this, but it's reasonable to conclude that he does not think the section to be firmly established and up-to-date.

The situation on the stability of the rings of Saturn is still somewhat confused. But important things have been done on the subject since 1833 so it is convenient to include the problem in the next section.

6.3   New results

6.3.1   *The rings of Saturn*

In 1851 Otto Struve, the son of Freidrich Wilhelm Struve and his successor at the Dorpat Observatory as master of double stars, turned his attention to the rings of Saturn. A new one, a faint ring located inside the others, had been discovered recently. The fact that none of the great observers of the previous generation had seen it raised the question of whether it had actually existed then, and he wanted to see whether the system overall was changing with time. He carefully measured the dimensions of all the rings on each side of the planet and concluded, "The smallness of the differences that result in general compared to the probable errors, makes us conclude, that at the time of these measures, Saturn appears exactly in the center of its system of rings" (Struve, 1851). Noting the careful qualification ("at the time of these measures"), we are still left with a negative result for offcenter rings. In the twenty-three years since Freidrich Struve's observation no confirming measures have appeared. What had been at the time a marginal observation is now a very doubtful one, and we can expect Herschel to take note and either add some words of caution to his account of it or leave it out altogether.

The matter of stability calculations has progressed also. In 1857 James Clerk Maxwell, as his entry in the Adams Prize Essay competition sponsored by the University of Cambridge, conducted a thorough and rigorous examination of the stabilty of Saturn's rings (Maxwell, 1857). The terms of the competition allowed him a choice of whether he might consider them solid, liquid, gas or an incoherent cloud of something. He looked at all these possibilities. At the beginning, he pointed out

> ...the cohesion of the parts of so large a body can have no effect whatsoever on its motions, though it were made of the most rigid material known on Earth (Maxwell (1857) pp. 286-7).

That is, the strength of a solid hoop of planetary dimensions is irrelevant. Both Laplace and Herschel mentioned the same fact, but neither followed it up in all implications.

Maxwell went on to do a stability analysis of Saturn's rings in several possible situations: solid and rigid (in spite of what he pointed out about rigidity on this scale), liquid or, finally, made up of separate chunks. In each he subjected them (mathematically) to all possible forms of small perturbation and worked out their response. His results were:

> The result of this theory of a rigid ring shows not only that a perfectly uniform ring cannot revolve permanently about a planet, but the irregularity of a permanently revolving ring must be a very observable quantity, the distance between the centre of the ring and the centre of gravity being between 0.8158 and 0.8279 of the radius (Maxwell (1857) p. 359).

That is, depending on details of the shape of the ring, it would be very obviously offcenter; this is well beyond anything allowed by observation. So Saturn's rings are not rigid solids. As for other alternatives, Maxwell concludes

> It appears, therefore, that a ring composed of a continuous liquid mass cannot revolve about a central body without being broken up, but that the parts of such a broken ring may, under certain conditions, form a permanent ring of satellites. (Maxwell (1857) p. 345)

Even Laplace's fluid rings are ruled out. So Saturn's rings must be made up of innumerable tiny pieces, each a satellite of the planet in its own orbit. Maxwell's conclusions supersede Laplace's analysis and makes the observation of a tiny eccentricity irrelevant to any idea of stability.

With this background we turn to Herschel's 1869 account of Saturn's rings. He starts out by conceding that there are observational reasons to believe the ring (here he retains the singular) is not solid, though in his wording they only pertain to the new, faint ring. Then he realizes that a solid ring of such size cannot retain its structure, as Maxwell noted in the beginning. "A fluid constitution would obviate this difficulty; and indeed it is very possible that the rings [note the plural] may be gaseous, or rather such a mixture of gas and vapor as consists with our idea of a cloud" (p. 434, §518). So he appears to have abandoned his earlier assumption that the ring must be solid.

In the next section, however, the original exposition of the solid ring is repeated exactly as before: there is the offcenter observation, the Laplace stability analysis, even the balancing pole. In light of the developments since 1833 it should have been heavily modified, if not omitted altogether; but nothing of the sort has been done.

While Herschel has included some recent observations of the rings, as a whole his section on Saturn has not been well updated and in fact is missing important developments. Certainly there was time to assimilate Struve's observation of 1851 and Maxwell's calculations of 1857; Herschel did not do so. I suspect (this is only a guess) that Saturn was just not among his interests, and he did not pay much attention to this section.

From the standpoint of the reader, Herschel's account of the rings is even more confused than in 1833. For several reasons they are probably not solid, but their stability is analyzed as if they were, using the same incomplete arguments as before. Herschel is grammatically inconsistent about whether they are singular or plural. If one is attentive, these are clues that the exposition is not entirely reliable.

6.3.2   *Stellar parallax*

As of 1833 no star had had its annual parallax measured. But within a decade, as Airy noted (see Section 4.2), at least one good result had been produced, while more followed in time for the Tenth Edition of the *Outlines*. Still, Herschel retains his earlier assertion that, "We know nothing, or next to nothing," about the distances, sizes or surface brightnesses of stars (p. 694, §780). "Nothing" is, strictly speaking, no longer true about distances. But with only a handful of stars out of tens or hundreds of thousands having their distances known, "next to nothing" is a reasonable thing to say. And consider that the parallax of the nearest stars is not much different from the upper limit of 1" mentioned in the *Treatise*, so almost all the statements made there can be taken over into the later edition unchanged; only here and there does the assumed lower limit of distances (a few light years) need to be changed into a real set of smallest distances. In a paradoxical way, the enormously important measurement of stellar parallax, sought for hundreds of years, has very little impact on Herschel's book. In part this is a historical accident, the assumption forced upon him in 1833 being not far from the real situation. Mostly this is because the overwhelming mass of stars remained just what they were, too far away to say anything about definitely.

6.3.3 *Thermodynamics and spectra*

Herschel is very wary about the new science of thermodynamics. His description of it is notable for its hesitation and doubt:

> A very old doctrine, advocated on grounds anything rather than reasonable or even plausible by Bacon, but afterward worked into a circumstantial and elaborate theory by the elder Seguin ... has of late been put forward into great prominence by Messers. Mayer and Joule and Sir W. Thomson ... Granting a few postulates (not very easy of conception, and still less so of admission when conceived), this theory is not without its plausibility ... (pp. 815-6, §905a)

This is about as faint a recommendation for a theory as one could look for. But it is entirely understandable. At this point thermodynamics was still incomplete (Helmholtz's presentation, Helmholtz (1863), has no mention of the Second Law) and its ideas still somewhat confused. The theory is introduced in the context of an attempt to explain the heat source of the Sun by infalling meteors, an explanation Herschel seriously doubts (for good reason); and so in a context not likely to recommend it to him. On the other hand, Herschel has himself tried a thermodynamic calculation of the difference in summer temperature between the Earth's northern and southern hemispheres, as noted above, though he does not yet grasp in detail the difference between heat and temperature, and certainly does not understand how complex the problem actually is. He deserves some credit for seeing in the new science the possibility of interesting and useful results.

More obviously useful, but even more engimatic, is the new technique of spectral analysis. No one knows *how* it works. The fact that a certain element would give dark lines in a spectrum when illuminated from behind, then bright lines in exactly the same positions when itself luminous, seems "so enigmatical as almost to appear self-contradictory," but "the conclusion ... seems inevitable." (p. 871, Note M) The possibility of identifying the very elements that make up celestial objects is exciting. Even more information is possible; Herschel describes in detail how William Huggins, by way of the Doppler effect, has measured the velocity of Sirius with respect to the Sun, some 29 miles per second (pp. 860-2, §859bb). Conclusions regarding the structure and possible power source of the Sun are less direct, the relevant passage (pp. 331-2, §400, n33) having been only slightly modified since 1833. But there are new and wonderful observations of the Sun being made (pp. 881-7, §395bb, 395cc, 395dd), so that "We have a new chapter of solar physics opened out, the commencement, doubtless, of a series of grand discoveries as to the nature and constitution of the great central body of our system."

Already, though, spectrum analysis has made a decisive contribution to the old question of the nature of the nebulae. That, however, we must consider in its place.

6.4   New problems

6.4.1   *Observations*

The increase in quality and quantity of observations during the decades separating Herschel's books has already been noted. These remain *visual* observations, made by people looking through telescopes and drawing or describing what they could see. As already noted, basing a theory or an explanation of something on just what it looks like has its dangers. There is also the possibility of an observation being simply mistaken.

6.4.1.1   *The nature of nebulae*

We have seen Herschel in his 1833 book being cautious about the nature of nebulae (see Section 3.4.2), noting that most seem to be formed of stars, and each increase of telescope power allows more of them to be resolved; but that some just don't *look* like masses of faint stars. It seemed probable to him that there was some kind of self-luminous fluid out there, as called

appears that no one understood quite how unreliable they might be until much later, when sensitive photographic plates allowed more objective tests. In addition, by this time Herschel himself was not observing regularly. He might have lost touch with that useful skepticism he displayed earlier, and perhaps have rated the power of the new generation of telescopes too highly; and he might have found the sheer mass of new observations impossible to analyze critically, and so been more inclined to take the observers at their word. With whatever explanation, among the reported observations are many that are inaccurate or mistaken.

There are observations and interpretations we cannot blame on Herschel's lack of current familiarity with the sky, however, because they are his own. One is the contention, emphasized to be a certain *fact*, that clouds tend to disperse under the full Moon (pp. 360-1, §432). It is ascribed to penetrating Moonshine being absorbed by the clouds, and is given some (very weak) support by rainfall statistics. This is probably an impression created by the fact that clearing clouds are more visible when there is a Moon around to show them (back when the skies were not well lighted by artificial means). In actuality, the light from the Moon is so much weaker than sunlight that it can have no detectable effect on the weather.

A larger question is the structure of the stellar universe, what we know of now as the Milky Way. It has been mentioned as a glowing band across the sky, resolved by telescopes into a myriad of faint stars. Sir William Herschel undertook a more thorough and quantitative exploration by carefully counting all the stars he could see in his telescope, in a large number of specific directions. As expected, there were more per square degree as he got closer to the central band of the Milky Way. If one accepts the two assumptions that stars do not vary greatly in intrinsic brightness (so that most of the difference we see is due to varying distances) and that they are distributed more or less uniformly per cubic light-year, the overall shape of the "universe" is as shown in Fig. 6.1.

Sir John had introduced his father's picture in his 1833 book, with the less than forceful recommendation that observations "agree with" the theory, a phrase repeated in this edition (p. 701, §786). By now further work had been done in counting stars, both by Sir John and by Struve, resulting in the conclusion that not only are stars found in a flattened distribution, but they are more concentrated as you get closer to the center of the distribution. That is, you will pass more stars by leaving the Sun and going a certain distance sideways than by leaving the Sun and going straight upward or at an angle; so that the assumption above of stars being

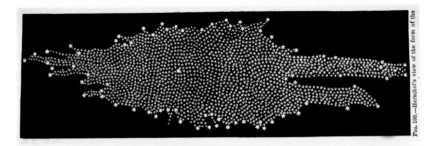

Fig. 6.1   A cross-section of the universe as derived from the star counts of Sir William Herschel. The Sun is marked by a larger star symbol at about the center and the stars of the Milky Way extend in a flattened structure left and right. To the right there is the Great Rift, visible as a bifurcation in the Milky Way many degrees in length. This diagram is taken from Newcomb (1878).

evenly distributed per cubic light-year needs to be modified (pp. 706-12, §792-6). Up to this point the results are in accord with what we now know: the system of stars that is the Milky Way is flattened, and the stars get more concentrated as one approaches its central plane.

Then Herschel points out (pp. 712-3, §797) that there is much variation from the averaged picture, with clouds of stars in some places and apparent voids in others. The dark area in the Southern Cross, known as the Coal Sack, is one of the latter (see Fig. 6.2). Herschel considers it to be a hole in a distant stratum of stars, because a void through many layers oriented directly toward our line of sight is so improbable (pp. 707-8, §792). In some places he sees very few of the faintest stars, and so concludes he is looking all the way through the Milky Way, past it to where there are no stars at all (p. 713, §797); in other places he sees only a single sheet, or two, of stars of all the same brightness; "The conclusion here seems equally evident that in such cases we look through two sidereal sheets separated by a starless interval." (p. 713, §797)

However, he does not insist on this categorically.

It is but right however to warn our readers that this conclusion has been controverted, and that by an authority not lightly to be put aside, on the ground of certain views taken by Olbers as to a defect of perfect transparency in the celestial spaces ... (p. 714, §798).

This takes us obliquely into cosmology, a subject we unfortunately will not deal with at any length, and Olbers' Paradox. That is: the night sky is dark. Why? If stars are uniformly distributed through space, then in any

Fig. 6.2   A part of the Milky Way, showing the stars of the Southern Cross and the dark area called the Coalsack (the section from the center toward the lower right). To a visual observer it could appear that there just aren't stars there; the impression is stronger in some other regions. This photograph is taken from Jeans (1929), originating from an amateur photographic atlas, the Franklin-Adams Chart.

direction you look you will, at some distance, eventually see the surface of a star, which is about as bright as that of the Sun; so the sky should be blinding. Alternatively, as you move out in distance from the Earth each star gets fainter, but there are more stars at that distance, the effects exactly balancing out, so again the sky should be bright.

Olbers thought the problem could be solved by some absorbing medium in space, so that more distant stars were dimmed by more than distance. (We now know that wouldn't work, but for reasons not known at the time of Olbers and Herschel.) Apparently Herschel was not completely happy about the origin of Olbers' solution, saying that support for the idea is "partly metaphysical." But another reason for his rejection of dark material is more interesting:

> We are not at liberty to argue that at one part of its circumference our view
> is limited by this sort of cosmical veil ... while at another ... star-strewn
> vistas *lie open* ... (Herschel's italics; p. 714, §798).

Well, why not? Herschel does not elaborate further, but the reasoning is not
difficult to reconstruct. If we can put this dark material wherever we want,
without any restriction, we can explain anything — which is just about the
same as explaining nothing. To put words in his mouth, if you allow any
distribution of magic dark stuff, you could use it to explain any difference
in brightness anywhere. Pushing it to the extreme, you could contend
that all stars are exactly the same brightness and at the same distance,
all the variation that we see being due to differences in the amount of this
obscuring matter. Unless there is some other way to detect it, some law by
which it operates, by invoking its existence you effectively stop any science
being done. It's a very powerful argument for a scientist, though it may
not be very clear why that's so, to a layperson.[4]

Returning to Herschel's interpretation of his observations, we find they
depend in particular on one assumption and its consequences. He is aware
of it: that stars are all of roughly equal brightness intrinsically, and hence
most of the variation we see is due to differences in distance. He has set
aside his earlier realization that, as far as he knows, they may differ among
themselves in the ratio of "many millions to one" (p. 694, §780), probably
because otherwise he could make no progress at all. But if they do differ
greatly in brightness, his conclusion that when he sees a "sheet" of stars of
the same apparent brightness, they lie all at similar distances, is untenable.
He is brought face to face with the problem by considering the extremely
varied population of the Large Magellanic Cloud, containing stars of all
apparent brightness (plus nebulae and clusters of all kinds) at the same
distance from us; "a conclusion which must inspire some degree of caution
in admitting, *as certain,* many of the consequences which have been rather
strongly dwelt upon in the foregoing pages." (His italics, p. 807, §894)

This assumption leads to another, which seems so plausible as not to
need stating: on the average, fainter stars are more distant than brighter
ones. We will look more closely at this in the next chapter. And further,
that there is a lower limit to the brightness of stars, so that if they're close
enough we can see the very faintest ones; and if everything we see is brighter
than the faintest possible, we have seen to the edge of the system of stars.

---

[4]There *is* such an absorbing medium in the Milky Way, interstellar dust; but the tools
to detect it were not developed for decades, and its existence wasn't generally accepted
until more than a half-century after Herschel's book. See Section 10.4.

But if stars have such a great variation in brightness among themselves, how is it that Herschel could see two distinct strata? He rightly points out that a variety of stars at one distance would certainly not look like a single sheet. There is also his conviction that in another place he could see a complete lack of the faintest stars. I can think of two factors that would contribute to the inaccuracy of these observations.

First, as Herschel would very freely admit, the precise measurement of the brightness of stars simply did not exist. The quantification of the ancient scale of magnitudes was different for each observer, and "... the unaided eye forms very different judgments of those proportions existing between bright lights, and those between feeble ones" (p. 695, §780). In other words, extending the scale to fainter objects by visual judgment is almost certain to be unreliable. His following pages and sections detail methods to nail down visual photometry, but in the end precision was obtained only with the advent of photography (and after a great deal of work on that medium). Herschel himself does not seem to have grasped just how inaccurate visual estimates of faint stars might be.

Second there is the matter of statistics. The distribution of star brightnesses in any particular field will vary randomly, depending on a number of things. Without knowing the areas to which Herschel is referring we cannot be sure, but it is possible that just by chance some striking configuration may form, and if someone (like the Herschels!) looks over enough of the sky the chances are good that the unlikely will eventually be discovered. Add to this imprecise photometry and the predilection of humans to see patterns even in random things, and I believe we can explain Herschel's strange observations.

Before we leave the Milky Way I want to bring up another matter of observation and analysis, convenient to mention here though more important a few chapters from now. It returns us to that strong point of nineteenth-century astronomy, the measurement of positions and changes in them.

I have mentioned proper motions, those changes in position peculiar to each star after things like precession and nutation have been accounted for. If we think of stars as moving individually, under the influence of gravity (whatever gravitational force happens to be at their position), we get a picture sort of like dust motes in a sunbeam, or leaves on a river; we expect each to move in at least a slightly different fashion, so proper motions are nothing unexpected. Indeed, compared to the average motion of all the surrounding stars, the Sun itself should be moving. We could only detect it by some systematic effect on the proper motions of every other star: they should be (on the average) spreading out from the place the Sun is heading toward, and closing in on the opposite side. Well, by the mid-nineteenth century this had been detected. Even though the Sun's motion makes up a small amount of what we see in other stars, if enough proper motions are analyzed it can be figured out. We saw Airy's evaluation of the first results in Section 4.2; the results as given by Hershel for the direction of Solar motion, set out in pp. 764-776, §852-862, are entirely consistent with the modern figures of Karttunen et al. (1996) within a few degrees. This should be astonishing considering the few data then available and the fact that almost no star distances were known. More astonishing is the fact that the rate (in kilometers per second) derived from these observations is of the right order of magnitude, since it depends on having a proper scale of distance.

The question of stellar motions among themselves raises the related one of general motions among the stars, that is whether there is an air current moving the dust motes or a general flow of the river. Herschel notes that the proper motion analyses to date would not have seen such a general motion (pp. 766-7, §853). But others have looked for something like an overall rotation of the Milky Way, in particular Johan Heinrich von Mädler. Mädler's result[5] gave a rotation about a position he identified with the Pleiades, which Herschel pronounces "improbable" since it means the whole Galaxy is rotating about a place well out of its plane (there are good dynamical reasons why this shouldn't happen). In fact he considers analyses of this kind "premature, though by no means to be discouraged as forerunners of something more decisive" (p. 775, §861). Mädler will be mentioned again in later chapters.

At this point I have to emphasize that most visual observations, as reported in the book, are accurate, and that they do lead to interpretations that have stood up over time. Take, for instance, Herschel's summary of what is known or inferred about comets (pp. 480-1, §570).

- First, the Sun excites matter in the nucleus into vapor, which escapes in the form of streams and jets and by reaction gives the nucleus an irregular motion;
- Second, this vaporization takes place on the side of the comet nearest the Sun;
- Third, some force from the Sun pushes the vaporized material away, forming the tail;

---

[5] Given in Mädler (1846), for those who can read old-fashioned technical German.

- Fourth, the Solar force (of an unknown nature) acts unequally on the vaporized material, some staying with the nucleus to form the coma, so there must be at least two types of material in the comet;
- Fifth, the Solar force is not gravitational, being repulsive and acting differently on different material;
- Sixth, the material in the tail must be lost to the comet entirely, since the nucleus does not have the gravitational pull to hold onto it;
- Seventh, the material lost by the comet must go into orbit around the Sun, in fact a family of orbits not much different from that of the parent comet.

All this is true as stated. The unknown Solar force is the Solar wind and radiation pressure, which act to push the tail away from the comet and has a different effect on the gases liberated from the nucleus than on the dust carried along. In fact the ionized gas liberated from the comet nucleus could very justifiably be called a form of matter entirely new to scientists of Herschel's time. Jets of gas which appear on the nucleus, caused by Solar heating, can push the nucleus in different directions (incidentally making predictions of the comet's orbit inaccurate). And the dusty material lost to comets does go into Solar orbit, sometimes intersecting the Earth's orbit in the form of meteor showers (which have in the past few years been predicted with great accuracy).

So visual observations are troublesome; interpretations based on them must be looked at with some suspicion; but we cannot simply throw them out as unreliable, because sometimes they are not only correct but also lead to the right answer.

6.5 Summary

With the tenth edition of his *Outlines of Astronomy*, Sir John Herschel has maintained his long effort of explaining the science to the most capable part of the lay audience. It is greatly expanded in size and coverage from his first effort, providing a detailed, wide-ranging picture filled with insight. Most of it is correct, and so of no direct interest to the present study. Similarly, much that is erroneous in hindsight is well-covered by caveats.

Several sections do show the strain of progress in astronomy, new results being patched onto old expositions without being completely assimilated. The reader does well to be confused here, and to maintain some doubt about what the true picture is. Examples are the stability of Saturn's rings, the resolvability of nebulae and the nature of the Sun.

More than in the 1833 edition, there are descriptions of disagreements among astronomers. Even when Herschel has a definite opinion as to which is the right side in a debate, the fact that he reports the opposing account means that it can be assumed to have some validity. Here again the reader should maintain a moderate level of doubt. Examples are the nature of the Milky Way as determined by star counts (where Herschel is mistaken) and the reasons for not seeing a central star in planetary nebulae (in which Herschel and Arago could both be considered right and wrong, but for reasons neither could know).

The major new problem with the 1869 edition lies in the evaluation and interpretation of visual observations. Herschel is simply too ready to report erroneous instances, such as the resolution of gaseous nebulae into stars, and to give too much weight to impressions, as in his interpretation of the distribution of faint stars in the Milky Way and in the tendency of clouds to avoid the Full Moon. At the same time, most of the observations are accurate and correctly interpreted; the picture of comets, in particular, is an example of how much can be done with only drawings and descriptions when the necessary physics has not yet been developed. So one cannot simply throw out all visual observations as unreliable. A true picture of how reliable they are did not really emerge until more objective methods of imaging and photometry were developed, and perhaps does not exist in detail even now.

The occasional unexamined assumption continues to make trouble. Here, the idea that all planets and the Sun must have a solid surface leads to some inaccuracies, even though it presents a clear problem in the latter case. On the other hand, the explicit assumption that there must be other forms of life in the universe, even intelligent life, does not seem to cause any trouble. It may not always be so innocuous.

Regarding our list of reasons for a scientist to make a mistake, again we have the troublesome evaluation of visual observations, made even more difficult when they're someone else's. Also, the erroneous observation supporting the questionable calculation occurs again. In the latter category we might put Herschel's notion of clouds disappearing under the full Moon. He had done much work in photography in different parts of the spectrum, showing that different colors affect chemicals differently. I speculate that an idea of the distinction between active and inactive moonlight, absorbed in layer of the atmosphere harboring clouds, might have made an impression seem factual.

Regarding our list of signs for the reader to harbor doubt about what he reads, we find the notion of a confused presentation reinforced. Indeed it may only stem from the almost unavoidable unevenness in updating a long-established book, when the science itself is changing, but in that case may also indicate that the author has not really assimilated new results. In general it might mean that the author himself is confused; I think the reader is entitled to assume the latter, if the writing is not clear, and maintain doubts about the result unless and until the author manages to clear them away.

And we may add:

- *Dispute or disagreement among scientists.* If an author reports with respect opinions on a scientific matter that differ from his own, it means that the matter has very probably not been conclusively settled. (This is *not* the same as describing Ptolemy's geocentric universe so as to show its flaws.) Reporting disagreement is another form of caveat or qualification, in which the author is demonstrating that, however firm his own conviction, the final answer may be otherwise.

Chapter 7

Simon Newcomb,
Popular Astronomy, 1878

The year is 1878. The United States has recently celebrated its centennial of independence, marking its evolution in ways much discussed by historians, but it is an appropriate date for our purposes here: many large telescopes are now located in this side of the Atlantic, together with skilled and industrious observers to use them, and our next author is a Canadian-born American.

It is less than a decade since Herschel's last book came out and, in the interval, nothing revolutionary has appeared in astronomy to shake up the science. However, the exploitation of techniques that were previously only of potential use can result in more real progress than the simple appearance of new things; we will see some of this in spectroscopy and thermodynamics.

Simon Newcomb was born in Canada and largely self-educated as far as science and mathematics went. In contrast with the dynastic beginnings of John Herschel and the prestigious institutional affiliation of George Airy, his beginnings can only be described as humble. But he has by this time become the Director of the Nautical Almanac Office, on the strength of his acknowleged mastery of the complicated and arcane art of computing the positions of the planets. He is thus the epitome of astronomy as *astrometry*, the traditional view of the science, as opposed to the new *astrophysics*, of which we have seen signs in Herschel's last book. He is also very much a mathematician and not originally an observer. From these facts we might expect very old-fashioned, highly technical, very theoretical books to come from his pen, even his *Popular Astronomy* (Newcomb, 1878) for the general public. That this expectation is not well borne out we shall see.

7.1 Audience and method

Who was Newcomb writing for? In words that sound rather familiar to us by now, in describing his book he says

> Its main object is to present the general reading public with a condensed view of the history, methods, and results of astronomical research, especially in those fields which are of most popular and philosophic interest at the present day, couched in such language as to be intelligible without mathematical study. (p. v)

Again we see that the division between popular and more advanced astronomy is marked by the use of mathematics. And, while in the previous books some mention has already been made of various bits of astronomical history, Newcomb intends to deal with it explicity; this important theme will be dealt with in more detail below. Finally, he notes that the readers will have their own reasons for being interested in the science, which he intends to take into account.

His disavowal of mathematics does not preclude, however, the occasional use of numbers. Where many an author would be content to note that the Moon does not have an atmosphere, Newcomb describes how a certain observation sets a numerical upper limit on its density (p. 315); in this he is like John Herschel. In his treatment of the theory of gravity he essentially translates algebraic formulae into words, then turns the results into numbers (pp. 77-8, 82), and expects his readers to understand what it means when something varies "inversely as the cube" (p. 90).

These might be simple slips of a highly-gifted mathematician, used to thinking in terms of formulae and numbers. His use of mathematical techniques, however disguised, is not nearly as prevalent or advanced as Herschel's. It still means that his audience is closer in capability to Herschel's numerate laymen than to Airy's "mechanicals" and working men.

I've noted that Newcomb was not himself an observer. He did, however, have to deal with many observations, and to estimate how reliable each one might be. We thus have the useful comment that 20" is a noticable difference (in position) between theory and observation, while 2' is embarassing, "not for a moment to be neglected" (p. 359). The declination of an object "can be determined to within two or three tenths of a second of arc, or even nearer, if all the apparatus is in the best order," but "When the astronomer comes to tenths of seconds, he has difficulties to contend with at every step" (p. 156). These numbers are useful to bear in mind when looking at observational results.

As noted, Newcomb brings in history explicitly. Indeed, the first three chapters, those covering from basic observations down to Newtonian gravity, are set in an explicit historical framework. It is refreshing, after Airy, to find an appreciation of the ancient Greek system as "a marvel of ingenuity and research when measured by the standard of its own times" (p. 33). And it is eerie, here in Chapter 7 of *Hindsight*, to find him proposing "to analyze these propositions of Ptolemy, to see what is true and what is false" (p. 35). He is critical of the failure to notice certain systematic features of the model that pointed to a heliocentric system (pp. 40, 58), but acknowledges that the lack of observed stellar parallax was a problem (p. 202).[1]

Even after the explicit historical part of the book, Newcomb continues to bring in history: accounts of the various observations of the transits of Venus, for example, and the similarly varied observations made at total solar eclipses. Along many lines of development he presents and discusses mistakes and discarded ideas in addition to those that turned out to be useful and accepted. Of course this approach is gratifying to a historian and adds some human interest to what might otherwise be a fairly dry scientific exposition; but its main importance, both to Newcomb's book and to this chapter, comes as part of the major theme of his book, to which we now turn.

7.2   Themes and content

7.2.1   *Progress and development*

Astronomy is the most ancient of the physical sciences, being distinguished among them by its slow and progressive development from the earliest ages until the present time. In no other science has each generation which advanced it been so much indebted to its predecessors for both the facts and the ideas necessary to make the advance. (p. 2)

This is how Newcomb begins his book, and the idea of astronomy as *progressive* runs through the whole volume. It is more completely expressed in a passage regarding the arrangement of what we now call our galaxy, the

---

[1] The alert reader will notice that, again, I am making a judgement on historical matters after I have explicitly denied expertise in the area. And, again, I will not be offended if readers do not choose to believe me, and go off to do their own research. I will point out that Newcomb has certainly read Copernicus (p. 60n), and goes into enough detail about the ancient systems to indicate that he has done his homework there also.

Milky Way:

> The preceding description of the view held by several generations of profound thinkers and observers respecting the arrangement of the visible universe furnishes an example of what we may call the evolution of scientific knowledge. Of no one of the great men whom we have mentioned can it be said that his views were absolutely and unqualifiedly erroneous, and of none can it be said that he reached the entire truth. Their attempts to solve the mystery which they saw before them were like those of a spectator to make out the exact structure of a great building which he sees at a distance in dim twilight. He first sees that the building is really there, and sketches out what he believes to be its outlines. As the light increases, he finds that his first outline bears but a rude resemblence to what now seems to be the real form, and he corrects it accordingly. In his first attempts to fill in the columns, plasters, windows and doors, he mistakes the darker shades between the columns for doors, and the pilasters for columns. Notwithstanding such mistakes, his representation is to a certain extent correct, and he will seldom fall into egregious error. The successive improvements in his sketch, from the first rough outline to the finished picture, do not consist in effacing at each step everything he has done, but in correcting it, and filling in the details.
>
> The progress of our knowledge of nature is generally of this character. (p. 478)

It is difficult to get across just how important this is, in several ways. If this book were a direct study of the history or philosophy of astronomy, I probably would have started here, with the idea of science as a *process*. It is implicit in John Herschel's assertion that there are ideas in astronomy which will not be overturned and very close to his exposition of "residual phenomema" (see Section 3.3), and it is probably safe to say that most practicing scientists look upon their profession in some similar way. Not all astronomers would agree that the development has been slow, and some would highlight the breakthroughs and mistakes instead of the average progress, but the emphasis on the process is generally agreed.

Newcomb's use of this guiding theme has important implications for the layman reading his book. First, there will be parts presented that are *known* to be wrong — in order to show the process. More subtly, there are parts expected to be wrong in some detail, though which detail is not yet known; this also responds to the reader's interest in "popular and philosophic" matters, the active areas of research that are not yet worked out. But the reader should be protected from complete error by Newcomb's awareness of the process as incomplete, mostly underlined with explicit caveats; and not least because the reader is reminded by Newcomb's use of the theme that he is looking at a process that, in the past, has thrown up red herrings.

In accordance with this theme, Newcomb mentions Huygens' method of estimating the distance to the Sun, which "may look like a happy mode of guessing" (p. 172) but demonstrates something of how to make rough estimates by using a reasonable assumption. It is, indeed, far more scientific than Mädler's orbit of the Sun around the Pleiades (which Herschel mentioned also; see Section 6.4.1.2), which is mere "baseless speculation" (p. 454). Possibly a bit more rigorous is the belief that the Sun is stable, based on the fact that most stars are of constant brightness along with "the general analogies of nature" (p. 435); the unlikelihood of detecting Earthlike intelligent life on another planet, given the brevity of civilization compared to the length of time the Earth has existed, is also "well-founded in analogy" (p. 519). I must emphasize that reasoning by analogy alone is very dangerous, and the cases here cited are actually a matter of probability plus a general assumption. They are thus somewhat better than "baseless speculation," though far from anything to trust in.

Newcomb makes the point that ideas may be partially or largely erroneous yet still useful. William Herschel's theory of nebular condensation was "almost purely speculative," and yet investigations in spectroscopy and thermodynamics since it was put forward indicate that "there must be some truth in it" (p. 452).[2] Indeed, there may be value in wrong answers: Professor Olmsted's ideas on the Leonid meteor shower, "Although . . . in many respects erroneous . . . were the means of suggesting the true theory to others" (p. 385). The reader is thus again made aware that steps along the process may give wrong answers.

And, indeed, the science itself (considered as the collective effort of astronomers) may accept wrong answers for an extended time, as in the instance of an incorrect value for the solar parallax (giving the distance to the Sun) for thirty years (p. 182). The important point is that wrong answers will eventually be corrected by the process.[3]

Newcomb's technique of setting out astronomy as a process and a development, in addition to implying that some of his statements will not be wholly true and simultaneously setting the reader on guard against that danger, also means he will at times disagree with other authorities and in many places explicitly warn against relying too much on what is presented. For instance, immediately after his analogy of an artist trying to draw a building in twilight given above, he says,

---

[2]The specific points cited by Newcomb in this passage are themselves in error, though his assertions are so hedged with caveats that they are not really misleading.

[3]Again I refer the interested reader to Trimble (2008) for how long "eventually" can turn out to be.

> But in the case now before us [the structure of the Milky Way], so great
> is the distance, so dim the light, and so slender our ideas of the principles
> on which the vast fabric is constructed, that we cannot pass beyond a few
> rough outlines. (p. 478)

In several places he shows a reluctance to rely entirely on authority. Concerning the construction of reflecting telescopes, for instance, he
presents an opinion of Mr. Grubb, "and there is no mechanician whose
opinion is entitled to greater confidence," but prefaces the remark with, "If
he is right" (p. 144). That this hesitation is warranted is demonstrated by
the fact that some of even William Herschel's observations were in error,
in particular his purported discovery of four additional moons of Uranus
(pp. 356-7). Newcomb's critical examination of observations in general is a
large subject, which we will get to a little later. But it is worth noting that
he is aware of the propagation of incorrect data from one writer to another,
and thus that a critical attitude towards published information is in order
(p. 341n on the rotation period of Saturn's rings, p. 376 on the supposed
Papal Bull against a comet).

In line with this distrust of authority as such are several accounts of
disagreement among authorities. For instance, mathematicians of mid-
century came up with different calculations and results for a certain minor
inequality in the Moon's orbit (amounting to 10 seconds of arc per century,
pp. 96–100). The details are inaccessible to a non-mathematical reader,
but the fact that the authorities disagree is significant, as is the fact that
the disagreement has not yet been resolved (p. 99). Newcomb mentions
that it might be some gravitational or tidal effect not yet accounted for,
including a variation in the length of the day (which technology could not
yet measure to the required accuracy), or a possible non-gravitational force.
(In fact the answer lies in the friction of the seas as the Moon's tide drags
them around the Earth, a difficult thing to calculate to any accuracy even
now.)

There is another theme that appears in Newcomb's book, not an important one but worth mentioning since it has appeared before. In this case
it is called the "plurality of worlds," in modern terms, extraterrestrial life
and intelligence. In general Newcomb is skeptical about the possibility of
such things in any given instance, commenting only in opposition to others
of less restrained imagination. The idea of living beings in the Sun is only a
"fancy" (p. 246), and on the far side of the Moon "no better than products
of a poetic imagination" (p. 309). That there may be some living on worlds
orbiting stars in a star cluster is given the faintest kind of support: "there is

nothing we know of to prevent" such a thing (p. 443). But "the attainment of any direct evidence of such life seems hopeless." (p. 516) Here in the twenty-first century there are many astronomers actively seeking extraterrestrial life and/or intelligence, so clearly much current thinking disagrees with Newcomb. But from the point of view of an astronomer three-quarters of the way through the nineteenth century the statement is perfectly correct; all current efforts use technology far in advance of any Newcomb could command.

7.2.2 *Content*

Newcomb's book is divided into four parts, each with subsidiary chapters. Here I list the number of pages in each part and chapter (the total in a part is larger than the sum of chapters due to introductory matter):

Chapter/Part	Title/content	Pages
Part I	The System of the World Historically Developed [The apparent sky through Newtonian gravity]	102
Ch. I	Apparent Motions	44
Ch. II	The Copernican System	23
Ch. III	Universal Gravitation	29
Part II	Practical Astronomy	129
Ch. I	The Telescope	40
Ch. II	The Application of the Telescope to Celestial Measurements	19
Ch. III	Measuring Distances [parallax]	45
Ch. IV	The Motion of Light	12
Part III	The Solar System	176
Ch. I	The General Structure	6
Ch. II	Inner Planets	48
Ch. III	Outer Planets	34
Ch. IV	Comets and Meteors	42
Part IV	The Stellar Universe	113
Ch. I.	The Stars [Variables, doubles, nebulae]	50
Ch. II.	The Structure of the Universe [the Milky Way]	31
Ch. III.	Cosmogony [star and planet formation]	29

In comparing Newcomb's book with those of Herschel (the subject matter of Airy was rather restricted by its format), the first thing I want to emphasize is how much they have in common. The great majority of the material in each addresses the same subjects, and it is indeed mildly surprising that (over almost a half-century) they do it much the same order and with roughly similar weights. More than anything else, there is a continuity in astronomy over the middle section of the nineteenth century, in spite of a great deal of change in the science.

Looking more closely, we find that the proportion of time spent on the planets, comets and meteors, and the Sun is almost the same between Herschel's books and Newcomb's. There is, however, much difference in detail; Herschel spends most of his chapter on the Sun describing observations, while Newcomb's emphasis is on the interpretation of observations and on theory (as we will see in depth a little later on).

Newcomb has significantly more (roughly double) in his description of Newtonian gravity, which is perhaps to be expected from the leading celestial mechanician of the time. But, ironically, he has *nothing* to match Herschel's exposition of the perturbations of planets, the core of Newcomb's professional work. (Perhaps he took more seriously the limitations of his readers, or since he worked with it at such a high level, thought it impossible to describe adequately.)

He also spends about three times as much proportional time on telescopes and instruments, reflecting something of their refinement since Herschel's time, and also the addition of the spectroscope to the astronomer's instrument kit. Much less time, about half, is spent on apparent motions, where he does not match Herschel's detail on the various proofs of the shape of the Earth (but does treat history explicity).

Newcomb spends almost twice as much time on the stellar universe, in part (another irony) going into much more detail about William Herschel's methods and conclusions about the shape and structure of the Milky Way than Sir John did.

Major new material in this book is found in his description of the constellations and in the star maps in the back, which amount to a basic observing guide. In addition, for each planet he mentions where one might look to find it in the next few years (after 1878). For all that Sir George Biddel Airy maintained over half a century before that someone could learn more by personal observations than by any lecture or book (Airy (1848), p. 88), a point with which none of our authors would seriously disagree, any help to make such observations has been thin on the ground in our books so far. Of course separate observing guides have been available, but this is the first time we've seen a star map (for instance) in the same volume as a general astronomy book written by an authority.

In summary, Newcomb avoids perturbations and spends less time on basics; more on history, telescopes and the stellar universe. Let us now look in detail at some further comparisons.

7.3   Old problems

With a new, younger author we expect a different approach to at least some of the problems identified previously, and here we are not disappointed.

7.3.1   *Saturn, craters and clouds*

We left the stability of Saturn's rings in a rather confused state, with Herschel still talking about offcenter observations and periodic perturbations of solid rings, while elsewhere considering that a fluid or gaseous state was more probable. Newcomb mentions Herschel's and Laplace's work, along with their defects as noted in previous chapters. He has assimilated Maxwell's 1857 analysis, and concludes that the rings are made up of a cloud of tiny satellites, "like separate little drops of water of which clouds and fog are composed, which, to our eyes, seem like solid masses;" at least, "This is the view of the constitution of the rings of Saturn now most generally adopted" (pp. 350-1). So, with a perhaps over-cautious caveat, Newcomb has disposed of this problem.

Concerning the origin of the craters on the Moon, Newcomb makes no mention of Herschel's "perfect volcanic character," but instead notes, "It is very curious that the figures of these inequalities in the lunar surface can be closely imitated by throwing pebbles upon the surface of some smooth plastic mass, as mud or mortar" (p. 314). He does not *say* that lunar craters are impact craters, merely notes a resemblance. This might have led to a serious examination at the two possible origins, but his treatment of the matter is brief and he doesn't compare the ideas. He does mention the rays, long streaks of light-colored matter that show up best at full Moon and seem to radiate from some craters. "The only way in which their formation has been accounted for is by supposing that in some former age immense fissures were formed in the lunar surface which were subsequently filled by an eruption of this white matter which forms the streaks" (p. 314).

change in the orbit of Mercury, a precessing of the perihelion to the amount of about 40 seconds of arc per century beyond what calculation could account for. He examines some possible explanations, but considers them unlikely, and leaves the subject open. With hindsight, we know that this comes from the fact that Newtonian gravity is not quite exact, and General Relativity is needed for measurements of this accuracy so near the Sun. That theory, however, is almost four decades in the future (though Newcomb has a very interesting hint of it; see Section 7.4.2), and Newcomb is quite correct in leaving it unexplained.

Note the size of the problems considered here: 10 and 40 seconds of arc in a century. The fact that astronomers could be worried about such quantities is a silent but powerful testimony to the increase in accuracy of both theory and observation over the middle of the century.

Overall, Newcomb says, "With the exceptions just described [the Moon's inequality and Mercury's precession], all the motions in the solar system, so far as known, agree perfectly with the results of the theory of gravitation" (p. 100). This is almost exactly the same statement made by Herschel forty-five years before. Newcomb, as Herschel, has not said that all is known about gravity, or any such categorical statement; just that everything (with two exceptions) fits to the accuracy of calculation and observation as achieved so far. Both statements are true as written.

7.3.3 *Visual observations*

As I've noted, Newcomb made his reputation as a mathematician, not as an observer. One might think that he would therefore be less critical of, certainly less effective in evaluating, observations. But remember that his specialty lay in using observations and comparing them with very exact calculations. He had to have a good idea of the accuracy of what he was given; that and his theme of progress and development made him, in fact, quite effective at evaluating the observations made by others.

We see, then, that the disappearance of the lunar crater Linnaeus, reported as such my Herschel, is dismissed as a trick of the light (p. 316), which in fact it was. The "willow-leaf" structure of the solar photosphere similarly is discarded (p. 238). Otto Struve's interpretation of observations of Saturn's rings as showing significant changes "has always been viewed with doubt by other astronomers" (p. 348), though the more general question of changes in other ways is not categorically rejected (p. 349). He shows a useful insight into how small mistakes made at (or near) the telescope can produce records of stars that aren't really there (p. 430), and how some "imperfection or optical illusion" can give rise to spurious times of rotation of Mercury or Venus very close to 24 hours (p. 293).

Indeed, Newcomb presents a useful general guide as to when some observations can be considered doubtful:

> There is a large class of recorded astronomical phenomena which are seen only by unskilful observers, with imperfect instruments, or under unfavorable circumstances. The fact that they are not seen by practiced observers with good instrument is sufficient proof that there is something wrong about them. (p. 287)

This doesn't cover everything, of course, but it's a good starting point. In all, then, Newcomb is a discriminating consumer of observations, and none that are worthy of doubt are presented in his book without some prominent words to that effect.

7.4   Progress and insights

With less than a decade gone since Herschel's last book, and no obvious spectacular theoretical results in that time, we might expect very little in the way of progress in astronomy for Newcomb to record. But, as noted, his is a new and younger set of eyes; and unspectacular progress is sometimes more productive. We can see some instances of the latter here.

7.4.1   *Spectroscopy and thermodynamics*

Spectroscopy, about which Herschel expressed some serious doubts, has now become an accepted tool of astronomy, enough that Newcomb devotes a full chapter to the spectroscope. No one yet knows just *how* spectroscopy works, but that it does is unquestioned; indeed, Newcomb expresses surprise that it went so long unused (pp. 222-3). The chemical composition of gases at any distance (so long as there's enough light) can be compared with gases on Earth, and familiar elements identified in the stars. The speed of a body toward or away from Earth can be measured by the Doppler effect, of which Newcomb comments, "The results of this wonderful and refined method of determining stellar motion, therefore, seem worthy of being received with some confidence so far as the general direction of the motion is concerned" (p. 458). But the measurements are not easy, and comparison of various results leads him to conclude that "numerical certainty is not yet attained." Similarly, although the temperature and pressure of a gas certainly affect its spectrum, there is as yet no way to determine their values directly from a spectrum (p. 229).

On a more practical level, it is very difficult to obtain the spectrum of a faint object. When the light is spread out into its component colors it becomes even fainter, placing a severe limit on what can be done with astronomical objects.

So while spectroscopy is now accepted as a part of astronomy, and has made a useful and important contribution to the science, more advances in theory and instrumentation (especially in fast photographic plates) will be necessary before it can fulfill its promise.

Thermodynamics, similarly, has made progress. Newcomb can state that, "it is now established that heat is only a certain form of motion" (p. 387), and that energy can be converted to heat and from it. (Recall Herschel's extreme doubts on the subject!) It can be used to eliminate certain theories of the Sun (p. 247) and, as will be seen, to provide a plausible power source for the stars. But the science of thermodynamics is as yet incomplete, having "merely empirical rules" (p. 281) relating heat and temperature, not useful at high temperatures (p. 241-2) or in the complicated situation of a mass contracting under gravity and radiating away some of its energy (p. 511). So while some results have been obtained, they must be viewed with doubt when presented in any detail, since aspects of the science which are as yet unknown in Newcomb's time may upset any particular set of ideas.

7.4.2 *Some insights*

A major problem with interpreting observations of the Sun (especially the outer parts), comets (especially the tails) and the Aurora Borealis in Newcomb's time was the lack of a theory of plasma, a state of matter of rarified, electrically charged particles. Plasma creates, and reacts with, electrical and magnetic fields in a very complicated way. The theory of electrical and magnetic fields was in place in 1878, but the idea of atomic or subatomic particles in space had not yet appeared.

However, I recommend a perusal of his pages on the Aurora Borealis (pp. 301-6). It is a very neat summary of what was known, including a number of observational clues unexplained at the time; what was unknown and unexplained; and what could reasonably be inferred. I present it as an example of how to treat a subject when you have a large body of observations, but still lack the physical theory required to come up with any kind of explanation.

I must bring out one other insight in Newcomb's book. In addressing what we would call cosmology, that is, what happens when you extend your knowledge of astronomy as far as you can think, he brings up the subject of non-Euclidean Geometry. In its modern form, this is curved space, and is central to Einstein's General Relativity (of 1915; see Chap. 10). Newcomb mentions that it is possible to have a space that is finite in extent, but has no boundary; in which a light ray proceeding straight ahead would eventually find its way back to its starting point (p. 505). These concepts, presented as pure speculation though they are, are a good forty or fifty years before their time. One wonders what Newcomb might have done with the full machinery of General Relativity, were he to begin his astronomical career fifty years later than he did.

7.5 New problems

We have seen Newcomb dispose of most of the problems (in the sense of this study) left over from Herschel. What new issues can he come up with?

7.5.1 *The Moon, Algol and some planets*

Nowadays, having celebrated the fortieth anniversary of the Apollo Moon landings, it is discouraging to be told that, "Human eyes will never behold the other side of the moon ... we are forever deprived of the view of the other side of our satellite," though Newcomb thinks it is probably just the same as the side we can see (p. 309). The latter point is arguably true while the former is, of course, wrong. But note, it is not strictly an *astronomical* statement; it is a prediction about technology. My inclination is to excuse Newcomb on that basis.

Similarly, as noted, Newcomb believes any search for extraterrestrial life and intelligence to be doomed to failure. While one can make probabilistic arguments about the possible number of inhabited planets in the universe (pp. 516-9), "Here we may give free rein to our imaginations, with the moral certainty that science will supply nothing tending either to prove or to disprove any of its fancies" (p. 519). Again, I hold that this is a technological statement, not an error in astronomy. More closely tied in with actual astronomical research is Newcomb's contention that "many centuries must elapse before we can do much more than tell how the nearer stars are situated in space. Indeed, we see as yet but little hope that an inhabitant of this planet will ever, from his own observations and those of his predecessors, be able to completely penetrate the mystery in which the structure and destiny of the cosmos are now enshrouded" (p. 409).

One can say that Newcomb should have been more aware of the possibility of surprises, having himself made the point that spectroscopy allowed certain things to be known that seemed forever beyond reach (p. 222); but this is perhaps a stretching of hindsight. The best statement to make may be a warning against any prediction of what may or may not be discovered, proven or calculated in the future.

On a much more mundane level, Newcomb rejects the idea that the variable star Algol is eclipsing, that is, that it gets fainter because one body passes in front of another, citing certain irregularities in the light curve. Well, Algol is in fact an eclipsing variable, quite regular. The irregularities cited must have been observational mistakes — quite an unusual error for Newcomb to make, and the source of which I have been unable to trace.

In the text, Newcomb states that Jupiter's density is less than water (p. 331). While it is less than Earth's, it is greater than water, as shown in his table in the appendix; I think this is simply a slip of the pen (Saturn would float on water; maybe he was thinking of the wrong planet). Since the actual correct number is available, I will count this as a typo and pass over it.

Regarding Uranus, I find one uncharacteristic and puzzling expression of certainty on the part of Newcomb. "No markings have ever been certainly seen on the disk, and therefore no changes which could be due to an axial rotation have ever been established; but it may be regarded as certain that it does rotate in the same plane in which the satellites revolve around it" (p. 354). Why is this certain? Since Newcomb's time observations have in fact established that Uranus does rotate in the same plane as its (major) satellites, but Newcomb explicitly has no indication of it — and yet regards it as "certain." Conclusions of much firmer character in the book are hedged with caveats and "probably." The fact that he was right in this instance does not make the statement, *in context*, any less erroneous.

7.5.2   *The Sun*

There are two areas that are amply supplied with caveats, so that there is no danger of the reader thinking that the answers given are certain or final. As such I would normally pass them by as not germane to our purpose. However, I think they are worth looking at in the conext of our overall question. This is because they shed some light on the question of *how much* of what an astronomer says should be trusted; they are extensive sections and the reader could certainly use some guidance as to consider them all speculation, or mostly well-established, or something in between. In addition, it appears that Newcomb (and the people he quotes and whose work he presents) were mistaken far more than they themselves thought possible. The first area concerns the Sun.

As I've noted, Newcomb spends a great deal of time on the power source and other theoretical (or at least interpretive) aspects of the Sun. This may be compared with two baffled pages in Herschel (1833), plus a short and very skeptical additional section in Herschel (1869). Newcomb is rather more certain he has *some* of the answers: "Respecting the physical consitution of the sun, there are some points which may be established with more or less certainty, but the subject is, for the most part, involved in doubt and obscurity" (p. 258). He proceeds to examine the corona, which might be (1) material constantly being thrown out at something like 200 miles per second; (2) particles suspended by electrical repulsion; or (3) tiny meteors in orbit, though "none of these explanations is much better than a conjecture, though it is quite probable that the facts of the case are divided somewhere among them" (p. 260-1). If pressed one might find the present understanding of the corona in some combination of (1) and (2), though they might be considered more lucky guesses than well-justified science.

Subsequently, he mentions the theory of Zöllner that the photosphere (the visible surface) of the Sun is solid, or a dense gas resting on a solid layer, as a way of explaining why it always has the same dimensions (which a gas presumably would not, under the great flow of heat from the interior).

The later part of the section on the Sun consists of four expositions by the experts of the day, four opinions about the physical state of the Sun and the processes important in it. They have quite different ideas about what sunspots are, all mistaken in hindsight; some are quite certain that "rain" and "snow" are important processes in the Sun, though not of water but of metals or other materials. Dr. Faye has fit a formula to the small motions of the sunspots (those apart from the general rotation of the Sun), "from which I conclude that we have to deal with a quite simple mechanical phenomenon" (p. 274). Even now sunspots are not completely understood, and are far from simple. He also claims better insight into rotating storms on Earth than meteorologists (p. 275). Prof. Young has doubts about Faye's theory of spots, "Still, it undoubtedly has important elements of truth" (p. 278). Prof. Langley is of a similar opinion.

My point is that, in spite of the caveats, all express far more confidence in their results than is warranted in hindsight, and this is after they have tried to separate what is well-established from what is speculation. In fact *most* of their conclusions are wrong. They are agreed, in addition, that the power source of the Sun is its contraction under gravity, which is not true.

In no way is this mass of error their fault. Physics in 1878 was decades away from any inkling of the nuclear reactions that actually power the Sun, and of any understanding of the motion of ions in strong magnetic fields; indeed, free ions themselves were unknown, and molecules and atoms a theoretical speculation. They were simply lacking absolutely vital tools to build an understanding of the star. They applied the science they knew to the observations they had. Their fault lay in vastly overestimating the progress they had made. They thought they understood a great part of the subject, when in fact they understood little, and misunderstood more.

Could this have been suspected? That is the important question. They did not know what the corona was, nor how it worked; theories of sunspots conflicted, and in any case did not quite match observations; they could not measure or calculate the temperature of the surface of the Sun, much less the interior. These are serious gaps. Perhaps most obviously (to a scientist, at any rate) is the fact that the rate of contraction cannot yet be calculated from the physics (p. 510); not only is it impossible to tell the Sun's past or future temperature, but the mechanism which has kept Earth at the same temperature over millions of years (the limits between which our seas neither freeze nor boil are, in a cosmic sense, very tight) is completely unclear.

I suggest, then, that progress in science is sometimes very detailed and extensive in the wrong direction; and, to put it in an overly simple form, if you don't know (nearly) everything, you may in fact know nothing. In any case, if an explanation contains any significant amount of "we don't know for sure" then the whole structure may be mistaken, regardless of how much confidence in it the authorities may have.

7.5.3 *The structure of the universe*

The matter of the structure of the stellar universe, what we now know of as the Milky Way, has already been mentioned; including Newcomb's very strong caveat (p. 409). It is certainly a bit unfair to criticize any of his conclusions as incorrect when he himself has said they are barely the first step in a process that will probably take centuries. But he believes that what has been done up to this point contains some truth, and constitutes progress (p. 478).

The picture he builds up, beginning (as far as any rigor is concerned) with William Herschel's star-counts, has the Milky Way a flattened stratum of stars surrounding the Sun; we are in the stratum, so that it thins away on each side of us, and close to the center of a generally circular layer (See Fig. 7.1). The irresolvable nebulae are found away from the center of the layer, for reasons that are not clear; star clusters tend to be found in or close to the layer.

As far as it goes, and without any regard to his hesitations, this is a reasonably accurate picture. We are close to the plane of the Milky Way, a flat galaxy of stars that thins out away from the plane. With

Fig. 7.1   The structure of the universe as figured out by astronomers by the second half of the nineteenth century, from Newcomb (1878). The Sun is about in the center of the picture.

no way of getting distances to any but the very closest stars it is a quite reasonable picture to build. I wish to point out, however, that it rests on two assumptions that in hindsight are mistaken. I believe Newcomb (and most astronomers of the time) were unaware just how important, and wrong, these assumptions were.

We have noted that Herschel discounted the possibility of some absorbing material in space, though Olbers had suggested its presence. Well, it is present. Dust, mostly in the plane of the Milky Way, dims light from most of the galaxy to the point that we cannot see it; and in fact we are well off-center, with most of the stars invisible. Newcomb does not even mention the possibility. For this book, the assumption of transparent space is that pernicious thing, an unexamined assumption. Given that he is aware of an absorbing atmosphere on the Sun (p. 240), he should at least have considered the possibility (if only to show why he didn't think it likely).

The other assumption is that, while stars may be of intrinsically different brightnesses, the range of brightness is not great: "in all reasonable probability, the diversity of absolute magnitude is far less than that of the apparent magnitude; so that a judgement founded on the latter [roughly constant brightness] is much better than none at all" (p. 461). On this assumption is built all of William Herschel's star-counting, as well as much subsequent work.

The most plausible definite formulation of this assumption that Newcomb gives is, "Although it is quite possible that an individual star of the fifth magnitude may be nearer to us than another of the fourth, yet we cannot doubt that the average distance of all the fifth-magnitude stars is greater than the average of those of the fourth magnitude, and greater, too, in a proportion admitting of a tolerably accurate numerican estimate" (p. 471). I am now going to demonstrate that this assumption can easily be completely wrong.

For this I am going to build a toy model of a part of the Milky Way. A "toy model" is often used in astronomy to describe a situation one sets up not as a serious representation of any observation, but to show the effect of some given thing. For this I will need to use definite numbers.

Suppose that, in a certain volume of space, there is one star of the zeroth magnitude; two of the first; four of the second; and so on. This is not an unreasonable sort of ratio, as we know now, since there are far more faint stars than bright ones. We take a chunk of space of just this size and place it, say, ten parsecs from us.

Now we take an identical bit of space with the same ratio between stars of a given brightness, and set it out at one hundred parsecs, that is ten times farther away. The stars will each be five magnitudes fainter (five magnitudes is a factor of 100; by the inverse-square law, something ten times farther away is 100 times fainter). But if we count stars in the same area on the sky as seen from Earth, the chunk of space more distant will actually have a hundred times the volume of the nearer chunk. So in the end, when we count the stars of a given magnitude on the sky, we get something like this:

Distance	0 mag	1 mag	2 mag	3 mag	4 mag	5 mag	6 mag
10 pc	1	2	4	8	16	32	64
100 pc	0	0	0	0	0	100	200

The average distance of a star of the fifth magnitude is just how many there are at 10 parsecs, times ten parsecs; plus how many at 100, times 100; divided by the number of stars. In this case the average distance is 78.18 parsecs. Do the same thing with the sixth-magnitude stars, and you get — 78.18 parsecs. *The average distances are identical.*

To show how much influence the form of the luminosity function has (that's the official name for the ratio of stars of different brightnesses), let's take a very different one: a flat function, with the same number of stars at all brightnesses; we'll say two, within the volume we'll set at ten parsecs. Building a table just as before, we get

Distance	0 mag	1 mag	2 mag	3 mag	4 mag	5 mag	6 mag
10 pc	2	2	2	2	2	2	2
100 pc	0	0	0	0	0	200	200

This time the average distance of a fifth-magnitude star is 99.11pc; of a sixth-magnitude star, 99.11. Again the same. The problem lies in the fact that there *are* stars in the near sheet that are of the same apparent brightness as the bright ones in the far sheet; that is, that the luminosity function extends over five (or more) magnitudes. Star-counts as performed in the nineteenth century just don't give any useful results. Now, Newcomb is aware that the range of absolute magnitude presents a problem:

> The differences of opinion which now exist respecting the probable arrange-
> ment and distance of the stars arise mainly from our uncertainty as to what
> is the probable range of absolute magnitude of the stars ... altogether, it ap-
> pears that the range of absolute brilliancy among the stars extends through

eight or ten magnitudes . . . It is this range of magnitude which really forms the greatest obstacle in the way of determining the arrangement of stars in space. (p. 483)

My point is that he is unaware that a range of this size makes any calculation based on the equal-luminosity assumption quite useless.

The problem of the luminosity function does not affect the qualitative picture of the universe as given by Newcomb. It is still correct in describing a flat system of stars, concentrated toward the central plane, in which lies the Sun, and out of which lie the nebulae. It does affect any calculation in detail, especially of the dimensions of this model.

As in the case of the physical condition of the Sun, Newcomb (and astronomers at large) are aware that their results are uncertain; but they are not aware (and especially in this case they should have been) just how uncertain they are.

7.6   Summary

Newcomb's *Popular Astronomy* is an excellent summary and presentation of astronomy as of 1878. He is more careful even than John Herschel about noting what parts of his material are uncertain, where there might be doubts about this or that conclusion. In fact as far as unflagged errors, one is almost certainly editorial (the density of Jupiter) and one is subtle, not to be revealed until Poincaré's work at the turn of the century (the stability of the Solar System). He corrects some of Herschel's mistakes (the full Moon dispersing clouds, the stability of Saturn's rings). And, remarkably for a non-observer, he is very accurate in evaluating others' observations; the only thing to be faulted is his acceptance of some phantom irregularity in timings of Algol.

The overwhelming value of his book, in relation to the previous ones, is his theme of the progress and development of astronomy over time. It is, I think, much more likely to give the lay reader an accurate view of the science, and a better idea of just how certain its results are, than any series of caveats or "probably" statements.

It is possibly in line with this theme that Newcomb presents the state of research on the Sun and the Milky Way with so much doubtful and unknown. No doubt the reader is made even more aware that astronomy progresses in stages, stages that include speculation and ideas that may be inaccurate or mistaken. But it is clear that Newcomb himself was not aware of just how erroneous these ideas might be.

Some previous items in our list of lessons have been reinforced. An immature science, as thermodynamics is still, should not be trusted very far, or in detail. Disagreement among authoritative scientists should prompt some doubt (regardless of how certain any of them might seem). Unexamined assumptions can be killers, and are very hard for the reader to detect. To these we add

- *If you don't know everything, you may not know anything.* This is put far too simply and neatly. The point is that, if your explanation or theory still has significant bits that are unknown, or that it simply cannot calculate, there is a chance the whole structure is wrong.

be inaccurate or mistaken. But it is clear that Newcomb himself was not aware of just how erroneous these ideas might be.

Some previous items in our list of lessons have been reinforced. An immature science, as the moderns like it still, should not be crushed too far, or in detail. Disagreement among authoritative scientists should warn us some doubt: (regardless of how certain any of them might seem). Unexamined assumptions can be fallacies, and are very hard for the reader to detect. To these we add:

• If you don't know everything, you may not know anything. This is put far too simply and nastily. The point is that, if your explanation or theory still has significant bits that are unknown, or that it simply cannot explain, then there is a chance the whole structure is wrong.

Chapter 8

Sir Robert S. Ball,
In the High Heavens, 1893

The year is 1893. It is the long afternoon of the *Pax Britannica*, and the advance of science and technology has continued in many areas. Steamships are the normal way of crossing oceans now, made rather more efficient through the science of thermodynamics, and the telegraph makes communication through most of the world nearly instantaneous.

Astronomy itself has advanced significantly without any obvious revolutionary developments. Telescopes are larger and there are more of them, now led by the large American refractors. Perhaps the most promising departure from practice to date is that some are built not where the astronomers are but where the sky is best for observing, in the American West and Southwest. More is being done with spectroscopy as a matter of routine and improvements in photography are making that technique ever more useful.

Back in England, the Lowndean Professor of Astronomy and Geometry at the University of Cambridge has brought out another of his books on astronomy for popular audiences, *In the High Heavens* (Ball, 1893). Sir Robert Ball is perhaps the least outstanding of the astronomers we shall look at as authors. He made no important contribution to the science on the scale of Herschel or Airy or even Newcomb: there is no Ball method of calculating planetary positions, no Ball catalog of celestial objects, no Ball equation. However, as a full professor of astronomy at the highly prestigious institution of Cambridge he certainly ranks as an authority for our purposes.

8.1   Form, purpose and readership

The book is a collection of "sketches of certain parts of astronomy which are now attracting a great deal of attention" (p. v), several of them versions of

143

articles contributed to the magazines *Contemporary* and *Fortnightly*. Previously published work, however, has been revised, and "in certain cases considerable alterations have been found necessary." We expect, then, no attempt at a complete coverage of astronomy on the form of Herschel or Newcomb, nor even a connected set of arguments such as Airy presented, though we may reasonably anticipate a level of consistency across all the chapters.

Ball's concentration on currently popular topics leads us to expect some excitement, but also a great deal of tentative presentation and simple "we don't know" since he will be operating on the edge of knowledge. Incomplete explanations should appear, similar to Newcomb's chapter on the Sun, and differing opinions should be anticipated.

The readership of a popular magazine in the 1890s could not be expected to reach the level of numeracy assumed by Herschel or (in effect) by Newcomb, and in fact Ball uses even arithmetic sparingly. To anticipate a bit, there are numbers employed, but more than anything else they are used to impress the reader with enormous sizes or distances or weights or whatever, rather than as a way of working out something. Ball's readers may be close to Airy's in background, but this book does not require their close attention to an extended argument as Airy's did.

There are no untranslated foreign languages, living or dead; the classical background of Herschel's readers of sixty years before is not shared by Ball's.

The book is divided into fifteen chapters, as shown in the following table:

Chapter	Title	Pages
I	The Movements of the Solar System	21
II	The Physical Condition of Other Worlds	20
III	The Wanderings of the North Pole	25
IV	The Great Eclipse of 1893	18
V	The Fifth Satellite of Jupiter	21
VI	Mars	32
VII	Points in Spectroscopic Astronomy	22
VIII	The New Astronomy	26
IX	The Boundaries of Astronomy	34
X	Is the Universe Infinite?	23
XI	How Long Can the Earth Sustain Life?	23
XII	The "Heat Wave" of 1892	18
XIII	Visitors from the Sky	39
XIV	The Origin of Meteorites	23
XV	The Constitution of Gases	17

There is no set of chapters on basic motions in the sky, nor an exposition of Newtonian gravity, explanation of telescopes, or other preparatory material such as we have seen before. This is as expected. There is, however, plenty on the planets, especially Mars, as well as aspects of the distribution and motion of stars (what I've called the "Milky Way" and what contemporary authors refer to as the "universe") and the nature of nebulae. In any particular chapter, indeed, we will at most see "aspects" of the subject, since Ball does not attempt any complete treatise.

8.1.1   *Themes*

Although the book is a collection of separate parts, there are a few themes running through the whole that can be discerned.

First, there is the awareness that great progress is being made through application and improvement of known techniques and principles, rather than through the invention of revolutionary new methods. The whole chapter on spectroscopy (VII) is a report of Huggins' work in applying the technique with increasing effectiveness over decades. In addition, "Already the camera has become an insidpensable adjunct in the observatory, and we are every day learning more and more of what it can do for us" (pp. 157-8); "We are only just beginning to realise the benefits from these photographic processes." (p. 167)

Next, there is the assumption that everything is perfectly adapted to its present environment; they probably wouldn't survive elsewhere, but are exactly fitted to be where they are. Humans probably could not survive away from the Earth, indeed away from their own particular place on it (p. 44); by a process that Ball leaves to one's own speculation, even the reduced gravity of the Moon would not "permit it to be an edurable abode" (p. 48). As in previous speculations about other forms of life in the universe, this assumption does not seem to lead directly to any identifiable wrong answers, though it might underlie some erroneous ideas, and it seems very strange to read that, "The planetary system now lives because it was an organism fitted for survival." (p. 224) The biological ideas of adaptation and survival of the fittest might be the same in effect as requiring the Solar System to be stable under perturbations, but one cannot avoid the possibility that the reasoning is wrong-headed in some way.

Finally, and most importantly, there is the idea that all warm or hot objects are steadily cooling down. Planets form an essentially one-dimensional sequence, placed at a certain point by their temperature, which is always decreasing; the smaller planets (which cool fastest) are in a state which the larger ones will reach later. The Sun itself is cooling inexorably (and might be on exactly the same sequence, at the larger end). The details of this theme will be presented and discussed a little later, but it is by far the most prevalent and consequential idea in the book.

From the organization and content of Ball's work, it should be clear that a detailed comparison with earlier books and problems would be difficult and probably not very useful. In addition, for reasons that will be easier to understand at the end than here at the beginning, I will not be looking so much at the *results* he presents but will concentrate on his *methods* of reasoning and exposition.

8.2   Logic and consistency

Ball begins the book with a chapter seeking to work out what the Earth's night sky looked like one million years ago. This is an interesting goal and one that admits of several definite conclusions, even with the limited observational and theoretical material available at the close of the nineteenth century. He begins with the closest star then known in the northern sky, 61 Cygni, and assigns it a typical speed with respect to the rest of 20 miles per second. Whatever its direction of travel, he concludes that a million years ago it must have been at least ten times farther away than it is now, thus a hundred times fainter to the eye; and, taking it as typical, concludes that *none* of the stars we now see would have been visible then.

Note that Ball has based his conclusion on the assumption that the *nearest* star is at a *typical* distance — which is nonsense. He was well aware that many stars are too far away for their distance to be measured, and that among these are some of the brightest. But he insists that none of the current stars could have been seen a million years ago, and thus "No methods known to us, or conceivable by us, can ever reproduce what the heavens must have been like at periods of millions of years ago." (pp. 20-1)[1] At this point I am not so much interested in the accuracy of Bell's distance for 61 Cygni or his assumption about its velocity, as in his glaring lapse in logic. It is not his last.

---

[1] Compare Tomkin (1998), in which the brightest star in Earth's sky is identified between four million years in the past and the same time in the future. The article is based on data unavailable to Ball, of course, but the techniques were known.

Another example of Ball's methods also comes from the motions of stars within the Milky Way. Recall that much earlier, indeed in time to rate a disbelieving comment from Herschel (see Sect. 6.4.1.2) and a disparaging one from Newcomb (see Sect. 7.2.1), Mädler had concluded from an analysis of proper motions that all the stars were rotating about one of the Pleiades. As we've seen, Newcomb did not believe it, and neither did Herschel; but we are not now concerned with its truth or plausibility. Instead, I want to look at how Ball goes about disproving the idea.

First, he *assumes* that the whole volume of the observable universe (that is, from Earth to the farthest visible object in all directions) "must be but a speck when compared with the space which contains it" (p. 233). Then he assumes that Mädler's rotation was meant to apply to the whole volume Ball has now imagined; and concludes that such a thing is so improbable as to be "preposterous."

There are two problems with his reasoning here: first, he constructs of Mädler's theory (rotation of the visible stars about a definite place) a sort of caricature (rotation of a vastly larger volume about the same place), and thinks he has refuted the one by refuting the other; but he has demolished a strawman, not the real thing. Second, he has assumed something for which he has no proof (and indeed, by definition, cannot have any proof), and used its assumed characteristics to reach his conclusion. With perhaps one intermediate step, he has thus assumed the thing he was trying to prove. Whatever the merit of the original theory, this method of disproof is wholly fallacious.

Circular reasoning, or something approaching it, appears in other places. In the chapter entitled, "Is the Universe Finite?" he asserts, "Adopting the sound principle that we need not assume more than is necessary to explain the phenomena actually presented to our consciousness, it seems to me to be clear that the number of molecules of matter in the universe must be finite" (p. 248). In other words, he will assume that all that there is in the universe, is what we can see (or, perhaps, infer from other means). In effect, he has assumed that the universe is finite and uses this assumption to prove it. Note, also, that this assumption is completely contrary to his assumption in Mädler's case.

Similarly, "showing" that differences in air pressure are caused by the Sun amounts to no more than the assertion that if the Sun went out there would be no more winds (p. 287). It may not quite constitute circular reasoning, but at any rate is perilously thin. And he confuses an *interpretation* of certain iron-rich stones in terms of his theory of the origin of meteorites, with *evidence* for his theory (pp. 351-3). It's the next best thing to assuming the theory, then citing it as evidence in support.

I've pointed out the inconsistency of Ball's assumptions regarding the universe beyond what is directly observable. Such inconsistency happens several other times. Presenting a calculation by Simon Newcomb, he asserts that the star Goombridge 1830, traveling through space at some 200 miles per second, cannot be slowed down by the total mass of all the estimated number of stars in the Milky Way. To be gravitationally bound to the system, there must be 64 times as many stars as are visible. "We are thus led to the conclusion that our system ... [is] visited by other bodies coming from the remotest regions of space" (p. 212). But, elsewhere, he asserts that, "even within the sphere which contains the visible stars that we know, there is such a stupendous quantity of matter of a dark character, that the visible part bears an almost imperceptible proportion to it" (p. 241). We will hear more about this dark matter later; for now, I only note that for one chapter it is ignored, for another it is dominant.

Similarly, he presents an argument to show that hydrogen is too light a gas to have been retained in Earth's atmosphere (p. 129-131, by the method I mentioned in Sec. 5.7); but an even lighter gas may be responsible for the aurora (pp. 163-4). In some places we are told that Mars has "oceans and continents" (p. 42) and "newly-discovered lakes" (p. 100); in another, that such things are unproven (p. 140). The latter may be put down to poor editing, but we are entitled to something better in a book that has seen "considerable alterations" in some chapters.

Finally, I will note an inconsistency regarding Ball's theory of the origin of meteorites. He held them to be produced by volcanoes on Earth, in the past when a much hotter planet had more powerful volcanoes than now; and not by volcanoes on the Moon, even though it would take less extreme activity to throw rocks hard enough to escape lunar gravity. To prove his point, he cites the fact that lunar volcanoes are all extinct, and so cannot be a source for any current meteorites (p. 321); but they could come from a "mighty primeval volcano on this earth" (p. 348). The point to note here is that the argument used against a lunar origin (there aren't any active volcanoes, much less of the power required) is simply ignored when it is just as applicable to a terrestrial origin.

These lapses of logic and consistency are bad enough on their own. In the context of Ball's book they are thrown into highlight by his occasional careful (one might almost say over-careful) reasoning. He notes that observations of binary stars are simply too rough to say they *prove* that

Newtonian gravity works the same way for them as it does for objects in the Solar System, though they are consistent with it and no one seriously doubts it (pp. 189, 197, 202-3).

So if Ball is willing to place that wonderful, precise machine of nineteenth-century Newtonian gravity into the category of "not conclusively proven," even if only formally, what does he truly rely on? There are two firm bases for his reasoning running through the book: the sequence of cooling objects and the myriads of dark stars.

8.3   Cooling planets and dark stars

"Suppose that you came into a room and found a pitcher of water on the table ... If you knew that the pitcher had stood there for an hour ... [and] If ... the water be in the slightest degree warmer than the air in the room, then the argument that it must have cooled from a higher temperature is irresistible" (pp. 32-3). It would be unfair to say that it took the whole of the nineteenth-century development of thermodynamics to reach the conclusion that hot things cool down, but certainly that sort of process forms most of the Second Law of Thermodynamics and scientists were given to making very general and far-reaching comments based on it. Ball extends it to the planets; since a larger planet will cool more slowly than a small one, it will resemble the smaller one at an earlier time, so "among these other worlds there are many in different stages, so to speak, of their development" (p. 32); in particular, Jupiter is hotter than the Earth (p. 97). The process also applies to the Sun, which will eventually turn solid and dark, indeed within the next ten million years (pp. 271-5). The more general form of this idea is expressed as, "We are bound to believe that heated bodies radiate their heat; and if so they must contract. This general law ... pervades all nature, so far as we know it ..." (p. 217). In fact it appears to be the only argument Ball will accept in favor of Sir William Herschel's theory of stars forming from nebulae, though elsewhere he appears to be certain of it (pp. 36-8).

Well, things *do* tend to cool down. But stating this process, as applied to stars and planets, as a general and immutable law, involves another assumption: "There is no means of replenishing to any large extent the heat of the inside of our earth by combustion. The earth's interior temperature must, therefore, on the whole, be simply falling in accordance with the laws of cooling" (p. 34). Elsewhere he is more brief and more general: "there is no source from which the loss [of heat by the Earth] can be replenished" (p. 261). This applies to the Sun also: "We have seen that it does not seem possible for any other source of heat to be available for replenishing the waning stores of the luminary" (p. 275). Here, "does not seem possible" is rather a weak way of expressing something on which the law of the cooling sequence depends absolutely, about which Ball expresses no doubt at all.

It is of course an unfair use of hindsight to fault Ball, or any other astronomer of the time, for not forseeing radioactivity (which helps warm the interior of the Earth) and nuclear reactions (which power the Sun). But he has made the entirely unwarranted step from, "we don't know of anything like this" to "there is nothing like this." It is true that combustion or a rain of meteors are inadequate power sources for the Earth or the Sun; it is *not* true that all other processes, including those unforseen, have been ruled out.

I will belabor the point a little more. It must often seem a simple step, especially to a scientist who has exhaustively studied a phenomenon and who knows that a great deal of intellectual energy has been spent in search of something to no avail, to go from, "nothing has been found" to "nothing is there." But it is unwarranted, and at the least such a jump leaves a hostage to any advances in the future. Now, there are instances in which we can say, "nothing is there." It is certain (for example) that there is no counter-Earth, orbiting always on the far side of the Sun from us and thus invisible; it would show itself through its gravitational effect on the other planets. But this kind of proof is rarer and harder to do than performing a search that turns up nothing. (It also tends to carry caveats, implied or expressed. The gravitational proof of the nonexistent counter-Earth, for example, actually means that nothing above a certain mass orbits there.)

The assumption that there is no heat source for astronomical bodies is not the only one of this kind that Ball makes. He also assumes that there is no source of oxygen on Earth, and so what exists in the atmosphere will eventually disappear (p. 137). This conclusion is not mentioned when he attempts to measure how long life on Earth will continue; but we have already seen that consistency across the book is not one of Ball's strong points.

It is a bit frustrating to examine Ball's theory of dark stars, because it is vague in detail and never presented directly. He is, however, very firmly convinced of it. "It is, however, plain that the ages during which the sun has been brilliant form only an incident, so to speak, in the infinite history of that quantity of matter of which the solar system is constituted"

(p. 242). Why it should be plain, Ball does not explain. "Every analogy would teach us that the dark and non-luminous bodies in the universe are far more numerous than the brilliant suns" (p. 274); but he presents no analogy in the section to make it clear. A bit of the workings of this theory (actually, to use the word "theory" is to give it a formal and definitive status that is, on evidence, unwarranted) is found in the comment, "It may be that the heat was originally imparted to the sun as the result of some great collision between two bodies which were both dark before the collision took place, so that, in fact, the two dark masses coalesced into a vast nebula from which the whole of our system has been evolved" (p. 275). This idea of collisions is presented rather tentatively, and in any case does not explain why there should be such dark objects in the first place.

The only support for this theory that is explicitly presented in the book is founded, Ball says, on the theory of probabilities. "By this theory we are assured, with a logic which cannot be controverted, that the invisible bodies must be vastly more numerous than the visible stars ... " (p. 241). He goes on to assert that the Sun, once it cools, will remain cool for an indefinitely long time; and that it must have been cool and dark for a great period of time before beginning to shine. Thus the period during which the Sun is luminous is tiny compared to the length of its existence, and the probability of finding something like the Sun shining is tiny; so most stars are dark (pp. 241-3). Indeed, he appears to believe it to be a general law, as strong as any in thermodynamics, that any bit of matter is only incandescent for a short period in its existence, and illustrates it by considering terrestrial iron (pp. 243-4).

Ball's logic, far from being incontrovertible, is circular. He assumes, on no evident grounds, that stars spend most of their lives dark; then assures us that the theory of probability proves his theory. Nonsense is almost too weak a word for what Ball has presented us with. His incontrovertible "dark stars" may have had more powerful support, in observation or theory, than he gives us (they could hardly have had less), but they have vanished from astronomy leaving no traces behind.[2]

Ball's confusion of speculation and assumption on the one hand with firmly-proven facts on the other stands out in higher relief because he has declared the intention to separate clearly "the truths that have been established in astronomy from those parts of the science which must be regarded

---

[2]At first glance, "dark stars" and what is now known as "dark matter" look alike, but the resemblence is accidental.

as more or less hypothetical" (p. 196). Like his attempts to be overcareful with logic, the effect is the opposite of his explicit program. This happens also in two major areas: his use of probability and his application of orbital dynamics.

8.4   Prerequisites for the course

Any scientist must have a good grounding in probability and statistics, astronomers perhaps more than most. In addition, the stock in trade of a nineteenth-century astronomer consists mostly of the orbits of objects in the Solar System (plus a few outside). Ball is fond of invoking the inexorable logic of the dynamical laws, as well as the incontrovertible power of the theory of probabilities (as we have seen). It is time to look more closely at his use of each of these.

8.4.1   *Probability*

Ball is concerned, early in Chapter X, with showing that the multiple star Theta Orionis is physically associated with the Orion Nebula, rather than being something at a very different distance seen by chance on the same line of sight. He finds that the chance of Theta being within the central square degree of the nebula is something like one in forty thousand (the number of square degrees on the whole sky), and thus a chance alignment can be discounted.

Well, the probability of Theta Orionis being in the same direction as the Orion Nebula is *not* one in forty thousand. It is *one*. There is exactly one Theta Orionis, and that's where it is. Ball has formulated the problem, at best, backwards; and since it's a common error (though one scientists should be well aware of), I'm going to look at it in detail.

Before you work out a problem in probability, you have to decide what is significant. This is a cardinal point. What would we accept as being a significant object to align with the central square degree of the Orion Nebula? Ball asserts, "there is only one star so marvellously complex in its character as Theta Orionis" (p. 239), which is simply wrong. There are many double and multiple stars in the sky, thousands and tens of thousands; even if we only chose those above a certain brightness, or multiplicity, there are hundreds at least. Add to that total the brighter star clusters, which would certainly look significant if projected onto the nebula, and Theta loses its special status.

(For the following section I'm going to work with some actual numbers, so if you're not comfortable with them just skip down a bit.) If there is one object, and we can place it at random anywhere on the sky, its probability of landing in any particular square degree (chosen beforehand) is indeed something like forty thousand to one (there are 41,252 and a fraction square degrees in the celestial sphere; the strange number comes from π being involved). If we have two, the probability of finding either in our chosen bit of sky is not quite twice that of finding one there (Consider the duck hunters of Sect. 5.6.) Working out some example numbers: if we have five hundred objects, and putting any one of them in the chosen square degree is significant, there's about a 1.2% probability of it happening by chance. That's not much, but it's far greater than one in forty thousand and should be enough to raise a doubt. If you decide that any of the naked-eye stars, about 6000 of them, is significant, the probability of it happening by chance is about 13.5%, which means it would not really be surprising if it happened or necessarily mean there is a connection. If you realize that both nebulae and multiple stars (and star clusters) are actually concentrated toward the Milky Way and not placed randomly on the sky, these probabilities increase (to work out numbers you'd have to decide just how concentrated they were).[3]

Suppose, though, you were still thinking about Theta Orionis itself, and wanted to limit your supply of objects to things just like it. You'd have to decided how much of a resemblence you needed, and if you insisted on something with exactly the same number of stars of the same brightness you'd be back where you started: there's only one Theta Orionis, and that's where it is.[4]

There are two points I want you to take away from this lesson. First, you have to set up a problem in probability very carefully, deciding what

---

[3]In all this I'm not suggesting that Theta Orionis is not associated with the Orion Nebula; in fact it is very firmly part of the object. But that is known from many other types of information, not an exercise in probability. My purpose in going through it is to show the error in Ball's technique.

[4]The problem of having only one example to work with bedevils many cosmologists and astrobiologists. We know of only one universe: this one; and only one planet with life on it: ours. Figuring any sort of probability means difficult decisions on what you would accept as significant, examining your work for unnoticed assumptions, and trying to figure out how your possibilities are distributed — the equivalent of how stars are distributed on the sky. It's not easy, and hard to be convincing.

is significant before you work the numbers. Second, what appears to be a negligable chance for one object can be quite a significant one for many, and "many" may not be a large number. Probabilities are counter-intuitive that way. If you have a group of 23 people, the probability is about even that at least two have the same birthday.

We have now looked at two instances of Ball's use of probability. In the case of Theta Orionis, his formulation of the problem involves an error (of a type that is fairly well known). In the case of his disproof of Mädler's theory of the rotation of the Milky Way, he made no calculation but simply decided (by construction) that a probability was preposterously small. In deciding that no meteorites could come from the Moon, he concludes again without calculation that the probability of an object from the Moon hitting the Earth "must be of a rarity so extraordinary, that it may be dismissed from consideration" (p. 320). In an exactly similar way he dismisses any chance of interstellar origin (p. 330) or a source on an asteroid like Ceres (p. 327). With no idea of how many objects might be sent from the Moon or an asteroid or the stars, any calculation of a total probability is meaningless. This is the case in most of his invocations of probability: he is satisfied if a superficial look at the situation allows him to characterize a chance as "very small," without actually dealing with numbers.

We have here an astronomer who does not attempt more than the most superficial exercise in probability, and gets that wrong; and yet claims "here again as elsewhere through astronomy the laws of probability afford a reliable guide" (p. 330).

8.4.2 *Orbital mechanics*

In dealing with the origin of meteorites, which have to come from outside the Earth's atmosphere, one necessarily deals with orbits. The trick is to identify a source, then a way for the rocks to get from there to here. Ball's answer is that volcanoes on Earth many years ago were far more powerful than they are now, and lofted rocks into the air with greater than escape speed, and he is concerned to show that the same mechanism on the Moon (for example), where the escape speed is significantly smaller, could not have provided a source.

Ball begins his work with orbital mechanics by asserting that a projectile sent upward from the Earth (neglecting, always, air drag) at seven miles per second would proceed at an ever-decreasing speed but eventually escape (pp. 314-6). This is fine, and in accordance with every analysis since Newton.

Now he follows a projectile sent outward from the Earth with this speed. He notes that, "In fact, long ere the little object had reached a distance which is as great as that between the moon and the earth, the sun's attraction would have surpassed the attraction of the earth" (pp. 345-6). Warned by Herschel (see Sect. 3.4.2), we note that this is neither surprising nor particularly important. But Ball goes on, "Assuming that the body moved in a straight line with a speed of seven miles a second, it is easy to show that after eight or nine hours it would have passed more under the influence of the sun than it was under that of the globe from which it had taken its rise. It is, therefore, obvious that ere long the object would be so far affiliated to the sun that it would have practically renounced all connection with the earth ... " (p. 346).

It is impossible for this body to move in a straight line at a constant speed of seven miles per second. Ball has either forgotten what he said fifty pages earlier (taking a full page to do so) about a body slowing down as it left the Earth, or has simply switched off Earth's gravity. In fact, by this point his body at escape speed has lost most of its speed relative to the Earth. But according to Ball somehow a body much closer to the Earth than the Moon is becomes able to ignore the gravity of the Earth, though the Moon is unable to do so.

The actual path taken by Ball's volcano-projectiles is not trivial to calculate, and would depend (for example) on which direction it was sent with respect to the Earth's orbital direction. You might look at Fig. 5.10, and consider what happens to rocks projected off the lunar surface at high speeds. If they are going fast enough to escape the Moon's gravity they may still remain in orbit around the Earth, forming a sort of halo of tiny natural satellites. Ball asserts that they will all have enough angular momentum that they will never actually hit the Earth (p. 320).

Next, look again at Fig. 5.11, and consider what happens to rocks sent off the Earth into an independent Solar orbit that eventually come back to the vicinity of the planet. Would they be more likely to enter Earth's atmosphere than their counterparts of lunar origin? It is a problem in three-body motion, of a sort that Herschel, Newcomb and Airy could certainly handle; on the evidence of Ball's exposition so far, I think it's reasonable to doubt his ability to do so. Similarly, it's fair to doubt his similar claim about the impossibility of an object reaching Earth from the orbit of Ceres (pp. 324-6).

We have Ball invoking powerful "laws of dynamics" as an agent of incontrovertible proof, then showing that he is incompetent in their actual use, as he has done already with the "laws of probability." This is terrible. He is a professor of astronomy at Cambridge, and hasn't shown the skill with probability or orbital mechanics required of a competent undergraduate.

8.5   Summary

Up to this point I could have recommended any of the books on popular astronomy as being useful and instructive, with minor reservations. Herschel was confused in places, but not in many, and not in important matters; Airy was perhaps too sure of himself, but his astronomy was sound; with Newcomb I was reduced to saying he knew less about the Sun than I believe he thought he did. Well, no longer. Half, perhaps a bit more, of Ball's book contains good and uncontentious stuff. That is simply not enough.

Not to put too fine a point on it, Ball's book is an embarassment. The conscientious astronomer could be forgiven for wishing books like this could be quickly and firmly supressed, or at least that they not be written by Cambridge professors.

I have not gone into any detail about what hindsight has to say about the truth or error of Ball's conclusions. I believe something more important is at stake here. You will recall that, last chapter, I identified Newcomb's theme of astronomy as a *process* as important. I want to emphasize that again. Indeed, most scientists I know will spontaneously and strongly tell you that any particular results they may get are *not* science, and are subject to later review, revision or discard; it is *how* they reach their conclusions that is science. Ball has written a book making a mockery of the process.

I think it's clear that the book falls far short of even everyday standards of logic and consistency, while claiming endlessly that this result or that is "plain" or "incontrovertible." He affects to distinguish between the well-established and the speculative, while placing absolute trust in things without foundation. Ball insists on enormous power and certainty for the methods of orbital mechanics and the theory of probabilities, while showing himself incompetent to use either. I don't think there's any question of an alert and attentive reader being misled by a mistaken presentation of something erroneous; the greater danger is that, seeing what passes for science in this book, such a reader will decide that that whole thing is complete nonsense and have none of it.

8.5.1 *Lessons*

What, then, can we salvage from the mess?

For scientists, of course we can demand logic, consistency, and at least a basic knowledge of the subject. There are two more advanced ideas:

- *Beware of attacking a strawman.* You do not disprove a theory or idea by disproving a caricature or oversimplified version of it. (This is a point well worth making in *any* argument or discussion.)
- *"We don't know of any" is not the same as "There isn't any."* You may eliminate all the effects, forces, reactions, sources that you know of; showing that there is no *possible* alternative is something else again. It may mean carrying around clumsy and inconvenient caveats (like "unless there is an energy source of which we have no inkling"), but it may save you from total disaster.

For the lay reader, as I've noted this books has many indications that it is to be taken with a great deal of doubt (if at all). Even aside from the problems with orbital mechanics and probability, which should probably be considered beyond the analysis of a true layman, there are enough gaps in logic and consistency to alarm an attentive reader. So I must reluctantly conclude that

- *An authority can produce nonsense, even in his own subject.* If there are flagrant gaps in logic, or the universe working one way in one chapter and the opposite in the next, you are entitled to doubt the results (or, better, find another author). Keep in mind that (especially when working at the frontier of knowledge) different assumptions may be tried out to see where they go, and lines of inquiry can be incompatible along the way. These, however, should be clearly signposted.

There are two other matters to mention:

- *If an astronomer claims to know more about a different subject than the specialists in the field, he's probably wrong.* We saw this last chapter with Dr. Faye and meteorology; Ball claims to understand geology better then the geologists (pp. 336-7).
- *Dismissal of counter-arguments by hand-waving is a sign of poor evidence.* Ball admits that "difficulties have been urged against" his theory of a terrestrial origin of meteorites; but, "I do not think I am over

sanguine in the belief that, serious as these difficulties may seem, they can be overcome" (p. 354). An appeal to some unspecified way of removing problems shows only that he has no idea how to do it, and underlines the possibility that it cannot be done at all.

Chapter 9

Simon Newcomb,
Astronomy for Everybody, 1902

The year is 1902. Although we have entered the 20th century and Queen Victoria is no longer with us, we are still very much in a 19th century world. Especially in the field of astronomy our theme continues to be one of continuous evolution, rather than revolution, though as I've noted with time our picture of the universe can change quite as completely this way.

So continuity and development is what we expect, especially as we return to Simon Newcomb as an author. Long-established in American astronomy now, he has come out with a new book on popular astronomy almost a quarter of a century after the first one we've looked at. *Astronomy for Everybody*, Newcomb (1902), has a slightly more colloquial title than his previous work and seems to have been popular for some time: my copy has photographs in it dating to 1925.[1]

9.1 The new book

9.1.1 *Motivation and audience*

The inspiration for this book seems to have been similar to Sir Robert Ball's, a series of well-received magazine articles. However, in Newcomb's case the articles only "suggested an exposition of the main facts of astronomy in the same style" (preface), rather than formed the bulk of the book itself. Thus we have an organized and (within its stated limits) complete picture of astronomy.

There is no reference to Newcomb's earlier work, and although there is some reused prose (for example, pp. 51-2, 110, 325) it is clear that the

---

[1] I have recently found a copy, revised by another author, printed in 1944 — thirty-five years after Newcomb's death. This was undoubtedly a popular book.

book is a fresh start on the problem. In organization and content this is a different book. We will not have to deal with the problem of fitting new ideas and results into chapters and sections written decades before.

The magazine-inspired style is simpler than the previous book, with shorter and simpler sentences. For instance, Newcomb introduces the Sun in this way:

> We see that the sun is a shining globe. The first questions to present themselves to us are about the size and distance of this globe. It is easy to state its size when we know its distance. We know by measurement, the angle subtended by the sun's diameter. If we draw two lines making this angle with each other, and continue them indefinitely through the celestial spaces, the diameter of the sun must be equal to to the distance apart of the lines at the distance of the sun. The exact determination is a very simple problem of trigonometry. (p. 91)

To understand the full difference in the style of prose, compare this with any of the quotations from Herschel's books. And also compare its complete lack of mathematics with Herschel's free use of trigonometry and algebra. (Just after this passage Newcomb gives some numbers, but any manipulation of them is still left out.) In the previous book, Newcomb had mentioned just how much of a discrepancy in Uranus' position led to the discovery of Neptune; here the number is absent. Physics is simpler also; the Doppler effect is presented, but no explanation is given (pp. 77-8).

At the same time, Newcomb does expect that his readers have seen spectra and know what they are (p. 49), and that he does not need to explain what solar and lunar eclipses are (p. 133; though he does spend some time on the phases of the Moon and how it rotates). Numbers appear when he compares various figures for the Solar parallax (and thus the distance to the Sun; p. 242), showing not only the numbers but how far different methods of determination agree, and he does expect his readers to understand squares and cubes (pp. 156, 250-1). In a more general way, he asks "the reader's careful attention," concerned that a superficial reading will not get his point across (p. 10).

His readership, therefore, is literate and remembers some of its arithmetic lessons, but is otherwise unskilled in mathematics and science, a slightly less capable set of people than his first book was meant for. Alternatively, he may be addressing the same people, but now they are relaxing with a magazine in the evening instead of more seriously trying to understand astronomy in detail.

9.1.2 *Content*

The book is divided into six chapters and many short sections, in this way:

Chapter	Subject	Pages
I	Celestial Motions	44
	A View of the Universe	6
	Aspects of the Heavens	10
	Time and Longitude	6
	Position of a Heavenly Body	6
	Annual Motion	16
II	Astronomical Instruments	40
	The Refracting Telescope	20
	The Reflecting Telescope	4
	The Photographic Telescope	2
	The Spectroscope	6
	Other Instruments	8
III	Sun, Earth and Moon	64
	The Sun	16
	The Earth	12
	The Moon	14
	Eclipses of the Moon	6
	Eclipses of the Sun	12
IV	The Planets and Their Satellites	104
	Orbits and Aspects	6
	Mercury	10
	Venus	10
	Mars	14
	Minor Planets	10
	Jupiter	12
	Saturn	12
	Uranus	6
	Neptune	6
	Measuring the Heavens	6
	Weighing the Planets	12
V	Comets and Meteoric Bodies	36
	Comets	22
	Meteors	14
VI	The Fixed Stars	42
	General Review	8
	Aspects of the Sky	6
	Constellations	16
	Distances	4
	Motions	4
	Compound and Variable Stars	4

The first point to make is that, compared with Newcomb's earlier book, this is much shorter: 333 pages in place of 528, and in larger type, so that there is maybe half as much material here. Next, the organization is different, so comparing how much space is devoted to each subject is sometimes inaccurate.

But we can conclude that about as much space *proportionately* is spent on apparent and actual motions, gravitation, telescopes and stars. Parallax receives a somewhat smaller proportion. Little or no space is given over to the motion of light, the structure of the Milky Way, cosmogony (what we would now divide into cosmology and star formation), and the details of theories of the Sun. In addition, the new book is not organized along historical lines nor to show astronomy as a process. There is proportionately (not absolutely) more on observations of the Solar System and on the fixed stars.

As in the earlier book, there is a section on the constellations and how to find objects in the sky.

The matter of other, complementary books has been mentioned, but it is worth repeating that nothing appeared in isolation. In this particular case, Newcomb notes that anyone seeking more details beyond his short section on the fixed stars in *Astronomy for Everybody* can turn to another of his books[2] (*The Stars, a Study of the Universe*, p. 291). The question of how a different depth of treatment affects this study's basic aim has thus been raised, and will be considered in this chapter.

9.1.3 *Summary of the new book*

Newcomb's new book is a smaller and simpler description of astronomy. It has less material in it and treats its subjects in less detail, especially avoiding the details of current research (the Sun, the structure of the Milky Way) where that is uncertain. The shift in content and emphasis is due almost entirely to the difference in audience and treatment, rather than the advance of the science in the intervening quarter-century. It is a new approach to the task, rather than a rewritten or re-edited version of the earlier book (for all that some of the prose has been reused).

---

[2]I'd like to remind you that full study of Popular Astronomy as a subject would take into account not only *all* the works by a given author, but those written by popularizers who were not really authorities on the subject as well as considering all the different levels of audience one might write for. It is far beyond anything I'm trying to do here.

From the standpoint of our exercise of hindsight, we may then expect fewer uncertain or contentious results to be presented, making fewer demands on the reader's judgement. At the same time, fewer details and supporting reasoning will appear, making it harder to develop doubts about the results even if doubts are really warranted.

9.2   Old, familiar problems

9.2.1   *Nebulae, the Moon and Algol*

The familiar question of the nature of the nebulae is still outstanding as Newcomb writes his new book. Posing the question as, "Are all unresolved nebulae actually made of stars?" We have seen Sir John Herschel undecided on the issue, then (under the influence of observations, some spurious, by the Earl of Rosse among others) almost certain the answer is "yes" until the spectroscope shows *some* of them to be certainly gaseous. Now the question has been refined to, "Are all the unresolved, non-gaseous nebulae made of stars?" Part of the difficulty of answering this question lies in the fact that nebulae are faint, thus very hard to get good spectra of. Especially when the spectrum is continuous, and thus spread out into all colors rather than concentrated in a few emission lines, nineteenth-century photography was very hard put to get anything at all. So few objects had good, detailed spectra recorded as the twentieth century opens.

In his first book Newcomb seemed to incline toward the idea that M31, the Great Nebula in Andromeda (just about the brightest nebula), showing a continuous spectrum but not at all resolvable, might be reflected starlight. The conceptual problem lies in the fact that a nebula actually made of stars that is not showing any sign of resolution must be very, very far away, and thus very, very big. It takes a bold scientist to assert that the observable universe is hundreds of times larger than everyone has thought, and of course there's a (quite justified) reluctance to embrace such a radical explanation before more mundane possibilities are ruled out. As Newcomb writes, it is fair to say that they haven't yet been ruled out.

In his first chapter Newcomb gives a picture of the Milky Way that is essentially the same as his first book, and as far as it goes is reasonably accurate; that is, the general shape and contents are right and the size is roughly correct. He asserts that there may be other things outside it, perhaps other patches of light such as the Galaxy might appear from an immense distance, "but of these we know nothing." That is, he does not identify them with any of the nebulae we can see. Arguing from omission, he makes the mistake of identifying what we now know to be external galaxies as objects inside the Milky Way. But this is a rather shadowy mistake, and it's just as easy to argue that by saying nothing on the matter he's made no mistake at all.

The origin of the craters of the Moon is another long-standing problem. In his previous book Newcomb mentioned *only* impacts as a possible explanation, somewhat surprising in light of the fact that everyone else seemed to be certain they were volcanoes. Now he does not even mention the impact theory, saying of the lunar craters, "There is, however, an almost exact resemblance between them and the craters of our great volcanoes" (p. 125). Note, however, he is not asserting this as completely proving a volcanic origin; strictly speaking he is only pointing out a resemblence. A bit later he is both weaker and stronger in support of volcanic craters: "Whether we accept this view or not, it is impossible to examine the surface of the moon without the conviction that in some former age it was the seat of great volcanic activity" (p. 126). The view can be accepted or not, and certainly much of the Moon *looks* volcanic (indeed, some of it is). A skeptical and alert reader would have been aware that the final answer is not in, though to me the abandonment of the impact idea without comment is worrying.

In the new book Newcomb corrects his dismissal of the eclipsing nature of Algol, removing one problem, and does not mention anything about the stability of the Solar System, avoiding another.

9.2.2 *Cooling and contracting*

We have seen how much trouble the theme of ever-cooling objects can cause in the wrong hands (meaning Professor Ball's). Newcomb has nothing more advanced to work with, but is more careful with it. He makes the point that the Sun cannot be solid or liquid, because its source of energy lies in contracting under the influence of gravity (pp. 101, 104-5); a correct conclusion based on an incorrect theory. The theory itself is "very probable" (p. 298). But its extension, that the Sun (and other stars) formed by contraction of a much larger mass (the "nebular hypothesis"), "is one on which opinions differ" (p. 106). Going further, "How did the nebula itself originate, and how did it begin to contract? This brings us to the boundary where science can propound a question but cannot answer it." In this book, instead of Ball's use of the hypothesis/theory as an unquestionable fact, and the speculations in Newcomb's earlier book, we have a simple, "we don't know."

9.3   Insights and progress

Throughout astrophysics, to this point, we have been bedeviled by the undeveloped state of thermodynamics. We have reached the point where the
energy gained from contraction under gravity can be related to the energy
radiating from the Sun (pp. 103-5), and so the Sun's lifetime calculated.
Radiant heat is related to visible light in some way, though (looking with
hindsight) there are many details left to be worked out (pp. 73-6). On the
other hand, we cannot yet calculate or measure reliably the temperature of
the surface of the Sun (p. 102). There is much left to be done in this field,
and Newcomb makes his reader aware of it.

As we saw in Newcomb's first book, although a mathematician by background and fame his judgement in evaluating observations is perhaps the
most accurate and reliable of our authors to date. In his new book this
continues. He refutes the impression that the southern sky (the sky as seen
from south of the Equator) is brighter than the northern, tellingly, by using
star-count numbers (p. 17). Let me emphasize this: a *visual impression* is
corrected by using *numerical measurements*.[3] He also recounts the different
results and disagreements concerning the rotation of Mercury and of Venus
(pp. 160-1, 169-72), concluding that probably no one has actually detected
either. He finds that the canals on Mars are not what some observers have
claimed, probably a combination of actual geographical features and what
we would now call features of the eye-brain system (pp. 180-5). In all this
he is correct. In the new book we find that he has done some observing
himself, at a solar eclipse (p. 172).

One other problem we have noted so far is the lack of a theory of plasma,
that is, the behavior of charged particles in the vacuum of space under the
influence of electric and magnetic fields. Without this understanding the
tails of comets were inexplicable (in spite of Sir John Herschel's strenuous
efforts to understand), to say nothing of the prominences of the Sun and
much else. Ball was led to contradictory conclusions about the composition
of Earth's upper atmosphere, though he did not notice the problem.

---

[3]Newcomb suggests that this impression is caused by the fact that travellers to the
south tend to visit areas that are dry (not cloudy) and free of industrial fumes, the Cape
of Good Hope for instance in contrast with industrial London. In my experience this is
probably part of it. But the Milky Way is definitely larger and brighter in the south,
something that would not show up in star counts unless one went to very faint stars, so
the impression is not entirely erroneous. The trick in this kind of comparison is to figure
out just what numbers correspond to what the eye is seeing.

In the new book Newcomb notes

Facts are now being discovered, and physical theories developed, the ultimate outcome of which may be an explanation of a number of mysterious phenomena associated with the earth and the universe. These phenomena are presented by the corona of the sun, the tails of comets, the aurora, terrestrial magnetism and its variations, nebulae, the Gegenschein, and the zodiacal light (p. 286).

The last three in his list did not turn out to be important examples of plasma physics, but with the tentative nature of his statement we cannot complain about them. He goes on to say,

The theories in question belong rather to the physicist than the astronomer, and the writer does not feel competent to explain them fully in their latest form, nor to define where established facts end and speculation begins.

In contrast with those astronomers who have confidently stepped outside their fields and sometimes challenged the experts, Newcomb makes clear that his understanding and exposition are possibly incomplete or inaccurate. But he goes on to show very well how a comet's tail can be explained in terms of ions and radiation from the Sun, not very far from the current picture; he even notes that the other things on his list are more complicated applications of plasma physics, which they are. His insight is very accurate as well as surrounded by the proper uncertainties.

9.4   Summary

In *Astronomy for Everybody* there is little to take issue with from the standpoint of hindsight, perhaps less than in any of the books so far. As far as the nature of the nebulae go, it is perhaps an error of omission. Some nebulae are certainly not resolved into stars but still lack a gaseous spectrum; Newcomb says nothing about them. We now know them to be external galaxies, like the Milky Way but very distant. In his picture of the universe at the opening of the book he leaves open the possibility of their existence, so it is hard to find a clear statement to take issue with.

The clearest example of an error in hindsight concerns the power source of the Sun. "The answer of modern science to this question is that the heat radiated from the sun is supplied by the contraction of size as heat is lost" (p. 104). Such a straightforward statement must be contrasted with his doubts concerning the nebular hypothesis, a sort of extension of the Kelvin-

Helmholtz contraction theory (and also, one might say, with his penetrating skepticism of visual observations by even experienced observers). From the book, the layman can derive no clue that the theory is in fact wrong. Can we derive any lesson from this?

To say that *any* particular statement in a book on popular astronomy may eventually turn out to be completely untrue is unsatisfying, and certainly isn't the case about (say) the fact that planets orbit the Sun, or the figures given (with uncertainties) for their masses. Perhaps the best way to look at this is to note that the contraction theory was new, cutting-edge science; and that is was incomplete: the details of how gravitational energy was converted into heat, and especially how it could maintain an almost unvarying output for millions of years, were unknown. Only scientists were aware of the second point (there is no mention of it in Newcomb's book), but it fits in with my earlier statement that if you don't know everything, you may not know anything. The first point should have been clear to a layman reading a popular astronomy book of the time.

Of previous problems, Newcomb corrected himself on Algol, avoids categorical statements on the craters of the Moon and says nothing about the stability of the Solar System. On the plus side, his evaluation of visual observations is excellent and his insight into electromagnetic explanations for things like comet tails and aurorae is very good. His shorter and simpler approach in this book compared to *Popular Astronomy* has led him to leave out much uncertain, current research, so while it is less exciting and detailed there is also less danger of setting out an incorrect theory. At the same time, he has not succumbed to the temptation of explaining things in so simple a manner as to be wrong, or to neglecting qualifications and caveats. It is a good example of writing for the less scientifically capable part of the popular audience without cutting corners. One would hope that it stimulated some readers, at least, to look for a deeper and more complicated work.

Chapter 10

Things Get Strange:
Quanta and Relativity

We have now well and truly reached the twentieth century, and physics is in the process of a revolution. Newtonian mechanics, which has served for every purpose over the past couple of centuries and more, is now known to be only an approximation to the truth. When speeds are a significant fraction of the speed of light it ceases to give the right answers. It also fails when objects the size of a molecule are examined, or when the interaction of light and matter are subject to calculation. The final direction of failure comes when gravity becomes very strong; but gravity rarely reaches that strength even among astronomical objects.

10.1 Relativity

Einstein's work with time and space appeared in two parts: the Special Theory of Relativity from 1905 and the General Theory of Relativity from 1915. As descriptions of their contents they can be misleading, especially to non-physicists, but "Special Relativity" and "General Relativity" are the way they're universally known.

10.1.1 *Special Relativity*

The first theory deals with what happens when speeds approach that of light. The heart of the problem is that many of the things that were taken for granted in Newtonian physics: the mass of a body, the measurement of distance, the flow of time; are now seen to depend on how fast an object is going. That is, how much mass *I* calculate *you* to have depends on how fast you are going relative to me; *your* clocks and rulers, *as I measure them*, will also seem to change. To you, they won't change at all — though mine will. In that way, "things are relative," in a precise and calculable way.

Another way to look at Special Relativity is through those things that do *not* change with speed. I will mention two kinds. The first is the speed of light.

First I have to remind you of your own intuition. If you're driving down the road at 50 miles per hour, and the car ahead of you is doing 60, you'll measure his speed relative to you as 10 miles per hour. Likewise, someone behind you who is doing 40 has a speed of minus 10 relative to you. It's a matter of addition or subtraction (in more complicated situations one would use vectors, but we won't need them).

But if I shine a beam of light ahead of me (which I measure as going at 299,792 kilometers per second) while doing 50 miles per hour, someone standing on the side of the road does *not* measure it as going 299,792 km/s plus 50 miles per hour; the stationary observer gets exactly the same result I do. In this case the discrepancy is tiny, but an equivalent experiment was done in the nineteenth century with speeds of hundreds of kilometers per second, and has been done since at even larger speeds. Light *always* goes at the same speed,[1] regardless of what you're doing.

The second relativistic invariant (a fancy way of saying "something that doesn't change between observers") will take a little more explanation, and again I want to remind you of your intuition. If you measure the length of an object, a pair of scissors or the edge of the table or whatever, you don't need to make sure what direction it's pointed. It will be the same length whether flat and oriented east-west, pointed directly up, or at any kind of angle. For things it's harder to slap a ruler against, we use the Pythagorean Theorem: take the length in each of the three spatial directions we have, square them, add, take the square root. The sum of the squares of the lengths in whatever directions you choose is an invariant in three-dimensional space.

Well, time is a dimension also. It's measured in a different way from space, but in relativity you can easily convert seconds to meters by multiplying by the speed of light (meters per second), which we know is a relativistic invariant. And you can set up the four-dimensional equivalent of a length, by bringing in also a duration; you choose a point at one time and another point at another time. But the relativistic invariant in this case, called the "proper time" or "proper length," is the sum of the squares of the spatial lengths *minus* the square of the duration (the latter, as necessary, multiplied by the speed of light). This is really strange. In three-dimensional space, if something has zero length then its length in every direction is zero;

---

[1] In a vacuum!

but you can have something over *here, now* and *there, later*, that is with a spatial length and a certain duration, with a proper length of nothing. (In fact it happens for two points in space-time that can be connected by a ray of light.) You can try to get back to a familiar kind of geometry to saying that the time axis is multiplied by i, the square root of -1, an imaginary number; but that interpretation is not now very popular, and leads to some problems with more advanced work in relativity.

There are things other than a simple interval/duration that can be set up in four dimensions, and have relativistic invariants. One of these is the momentum four-vector, the relativistic extension of the momentum-arrow we worked with in Chapter 2. Its time-component has the units of energy, and it is not a zero-length vector for any body that has mass. If we choose to travel along with the mass, it has no components of the four-momentum in any of the spatial dimensions, but it still must have energy. That is, we are slewing this four-vector around like a pair of scissors until it points in a convenient direction, in this case along the time axis. And we find that an object has energy by virtue of its mass, even when standing quite still. This is the source of Einstein's famous $E = mc^2$, giving the amount of energy contained in mass by its very existence.

Immediately, we must modify the science of thermodynamics. Energy can neither be created nor destroyed, only converted from one form to another, *one of which is mass*. I should point out that the formula gives no guide as to *how* one would go about doing such a conversion, only that it is theoretically possible.

Next, we learn to look for some change of mass in our energy-conversion reactions. So much coal and oxygen, together, should weigh more than the resultant carbon dioxide after we burn them. The difference is tiny (a very little mass makes for a lot of energy) so it would not be practical to try to measure it in this case. But when dealing with tiny things like atoms and their component parts, as we are in the twentieth century, the masses are small enough and the energies large enough for believable results to come in.

10.1.2 *General Relativity*

Special Relativity deals with motion at high speeds; General Relativity is a theory of gravity. GR reduces to SR if you take out all effects of gravity, so it's completely consistent with the earlier theory; but it's mathematically and physically a much richer field (which means far more difficult to do calculations in). It does not actually come into our use of hindsight much at all. Up to the middle of the (20th) century there was some work done in the field, but the observational consequences were tiny and it was not a major effort from the standpoint of astronomy as a whole. Nevertheless, it required a major shift in thinking about gravity, and it's worth outlining some of its features.

In GR, gravity is not a force, changing the momentum of bodies through its action on them. Gravity is a result of the curvature of space-time, through which bodies (if they're not subject to other forces) follow *geodesics*. All this needs explanation.

What is a straight line? Is it the shortest distance between two points, the way to draw a triangle so the angles add up to 180°, the way to draw a triangle so the Pythagorean formula holds, the way to measure a distance so that the familiar formulae for area and circumference of a circle come out right? In flat space, in Euclidean geometry, you can do all of these things at once. If the space is curved you cannot, and have to choose which features are most important. A geodesic is that path which follows the shortest distance between two points, if they're taken to be close enough.[2] It's best, again, to work with examples.

Since four-dimensional space-time is hard to think about, we'll move to a two-dimensional space: the surface of a sphere, like the Earth. Remember we're restricted to the surface, to moving in two dimensions, and we are not aware of, or capable of dealing with, the real third dimension. Geodesics on the Earth are "great circles:" if you slice the Earth so that your slicer passes through the center, the intersection of the slice with the surface of the Earth is a great circle. Meridian lines are great circles; lines of latitude, apart from the Equator, are small circles.

Suppose we send two people from the North Pole along two different lines of longitude (in the best tradition of mathematical physics, we smooth out all the features of interest on the globe and ignore things like wind and temperature). These are great cicles, so they're going along geodesics. They start out at some angle to each other and move apart; well and good.

But after a while they notice (through some kind of radio rangefinder or similar device) that they're not moving apart as quickly as they once were. Indeed, as they pass the Equator they're not moving apart at all, and start to come together. They crash into each other at the South Pole unless they

---

[2]This "close enough" comes from the fact that, if a space is positively curved, there may be more than one distinct path of the same length through two given points. I wrote a highly mathematical and obscure paper in this area once; most astrophysicists don't worry about it at all.

stop first.[3] If they considered the two-dimensional surface of the Earth to be flat, they would conclude that they've been subjected to a force pulling them together. This is how curved space can appear to exert a force.[4]

In General Relativity, mass curves space-time; masses move on geodesics in this curved space-time. The mathematics of how it's all done is formidable. The physical interpretation of the theory was still being worked out in the middle of the twentieth century, and one could say it's still not quite finished now.

How do you know your own three-dimensional slice of space-time is curved? Well, the various definitions of a straight line won't all agree. One way would be to construct a sphere, and measure the volume enclosed along with the surface area. If there is *more* volume than there should be, that is if you can stuff more in your suitcase than it should hold (think of many a cartoon from years ago), space is positively curved. Geodesics will bend toward each other; this comes from positive masses. If your sphere holds *less* than it should, space is negatively curved. As a practical matter, in almost all of space the curvature is so slight that one could not actually measure it this way, but it does provide a way to think about the effect.

There are three bits of observation available in the first part of the twentieth century in which General Relativity appears. When gravity is strong, it appears to grow even more quickly than the Newtonian inverse-square law. That modifies the orbits of planets slightly, in the case of Mercury giving rise to a shift in its perihelion by something less than an arc minute per century. This is tiny, but the precision of the instruments of the later nineteenth century measured it without any question, as we have seen; and it had been a puzzle.

Second, even light (which is massless) could feel gravity as a warping of space; so the light of stars passing close to the Sun is shifted a little. This amounts to less than two seconds of arc at best, close to the limits of that type of observation, but was confirmed at an eclipse in 1919.[5]

---

[3]Lines of longitude, connecting the North and South Poles, are examples of distinct paths with the same length; there is no path of "shortest length" between these two points. This is the motivation for my previous footnote.

[4]In GR you are concerned not only with the curvature in space, but also including time, which gets complicated because of that minus sign in the equivalent of the Pythagorean Theorem. We don't need to worry about the details here.

[5]Strictly speaking, Newtonian gravitation made no prediction for the effect of gravity on light, since the nature of light was unknown. There were ways of attributing some effective mass to a light beam within classical physics, but they gave a result which was less consistent with the 1919 observations than GR.

Third, light coming from a deep gravity well (that is, climbing away from a concentrated, massive body) would show something that looked like a Doppler shift, though always toward the red. For the Sun this was detected only marginally, but for a class of stars called White Dwarfs is quite clear and measurable. This "gravitational reshift" is probably the most important GR effect for our present use of hindsight.

10.2 Quantum Mechanics

For those who like to remember their history neatly, a paper read by Max Planck in 1900 makes a convenient starting date for the Quantum Theory. It took decades to work out in full, however, and astronomers were to use it in incomplete condition for many years. While its effects (somewhat surprisingly) could appear in very big phenomena, it is really a theory of what happens on very small scales.

10.2.1 *Some nuclear physics*

By the second quarter of the twentieth century, two particles smaller even than atoms had been discovered: the electron, with a negative electrical charge, and the proton, a positively-charged object with a mass something above 1800 times that of the electron. Both were present inside atoms; the protons in the tiny, massive nucleus, electrons running around outside in a region a thousand times larger (though maybe there were some of these in the nucleus also). The different types of atom came from different numbers of protons in the nucleus, giving different masses. A different number of protons also meant a different number of outside electrons to balance the electrical change as a whole, thus explaining the whole Periodic Table of the Elements.

Some very heavy elements were found to be radioactive. Uranium and radium, to take two examples, would emit very high-energy particles and radiation spontaneously, at the same time changing into lighter elements. This *fission* would occur at a specific rate: starting with a certain amount of the fresh radioactive element, after a period of time called the *half-life* one half of the material would have fissioned; after another half-life, one half of what was left; and so on. The radiation given off would thus initially be rather strong, then grow weaker and weaker over time. No changes in temperature or pressure or anything available in the laboratory appeared to make any difference in the rate of fission. Even stranger, there seemed to be no way to tell when a particular atom would fission: it might be right away, it might be thousands of years from now, all very puzzling.

With Einstein's famous equation in mind, it was worked out that four protons (each proton being the nucleus of a hydrogen atom) together weighed a trifle more than a helium nucleus (with some electrons thrown in where necessary to make the charges come out right), so perhaps such a *fusion* reaction could be used to power stars. It certainly would release more energy than the Kelvin-Helmholtz contraction process, which had been known to be inadequate for a long time. (Dating of rocks by radioactive decay showed that the Earth was several billion years old, so the Sun had to shine for longer than the few million years that contraction could last). Indeed, fusion was much more energetic per mass of material than fission. But no fusion reaction had actually been observed (it takes a lot of forcing to get two positively-charged nuclei close enough to interact). Of course even more powerful would be the total conversion of matter to energy, in which all the mass was used up, but how this might happen was at best a speculation.

10.2.2 The Old Quantum Theory

By the late 1920s a mostly coherent theory of small-scale physics had been built up. It was known to be incomplete, and in places gave contradictory answers or none at all, but was outstandingly successful in some areas where classical physics[6] was completely useless. For convenience we call this the Old Quantum Theory, as if there was a distinct pause at one point, though of course there was development occurring all the time.

10.2.2.1 *Waves to particles: blackbody radiation*

To introduce quantum mechanics it's useful to start with another of those ideal things that don't actually exist, but are very useful: the blackbody, with its associated blackbody radiation. A physicist describes a blackbody as something that absorbs all radiation completely, and thus radiates all

---

[6]The term "classical" physics is normally used, as I use it here, for Newtonian mechanics and Maxwell's electromagnetics; that is, for physics before the theories of Relativity and Quantum Mechanics. I have also seen it applied to General Relativity as distinguished from various quantum theories, which could be confusing, and sometimes one might find it used to mean "the older theory" as distinguished from any new development. I'll try to be consistent and precise, but you may find authors with other ideas.

radiation perfectly; which is not at first glance a particularly illuminating idea (no pun intended). Well, think of things that are *not* blackbodies: a rarified gas, for instance, which absorbs and radiates only in its particular spectral lines; or the green leaves of a tree, which reflect some of the green light that falls on them (instead of absorbing it). A colorless solid or liquid or gas at high pressure is a better approximation of a blackbody. To be specific, consider a black iron poker thrust into a very hot fire. It will absorb the radiation from the fire very well. At some point it will reradiate the energy it has absorbed, glowing red when it's hot enough, then yellow, then (if it gets very, very hot) blue-white. This reradiated energy comes out in a form that depends only on the temperature, not at all on what the blackbody is made of.

Another example of blackbody radiation is found in a cavity surrounded by a material at some particular temperature, when the radiation is in equilibrium with the material, that is, radiation is absorbed at just the same rate it is emitted. "Cavity radiation" is an equivalent name for "blackbody radiation," and in fact is the specific situation treated by Planck in 1900. (Sir James Jeans, two of whose books we'll look at, also uses the term "full radiator" in place of "blackbody.") This means that a chunk of electromagnetic radiation can be thought of, quite rigorously, as having a temperature.

The problem with blackbody radiation, from the point of view of classical physics, is that there is only so much of it. Calculating what the energy density of cavity radiation should be using just Maxwell's electromagnetics and the formulae of thermodynamics, one concludes that there should be an *infinite* energy density in the cavity, regardless of temperature. Also, calculation gave far more radiation at short wavelengths than anyone actually measured.

The solution to the problem lies in the discovery that the energy of electromagnetic radiation of a given wavelength comes only in discrete chunks, in units, so that one can have one unit of energy or fifteen but not six and a half. Now, light was known to show all the features of a wave, and the energy of a wave is continuously variable: it can take on all values within a range, just by varying the amplitude (see Sect. 5.1). Energy in separate units, on the other hand, is more like counting the mass of a bunch of identical particles — so if light is a wave, how can it be a particle? But it all works out that way. Doing the calculations, the spectrum of blackbody radiation comes out right (Fig. 10.1).[7] Light-energy comes in individual packets called *photons*, whose energy is a specific function of wavelength: the shorter the wavelength (higher frequency), the higher the energy of each one. Sir John Herschel's actinic rays are explained easily: red and green light photons just don't have enough energy to perform the chemical change required to make a photograph, while blue and ultraviolet photons do. If you shine a light of a frequency below a certain number on a metal in a vacuum, nothing happens, no matter how bright you make the light; once you increase the frequency past a critical value, electrons are ejected from the surface of the metal. Photons need enough energy individually to release an electron.[8]

I have spent so much time on blackbodies not so much for historical purposes as because they are so useful. Almost any solid, liquid or high-pressure gas behaves, to a first approximation, as a blackbody. This includes stars. We know stars have absorption and emission lines, so they're not perfect blackbodies, but for the purposes of working out some initial answers they're quite close enough. And since we can calculate the energy sent out per square meter of area from a blackbody, once we fit a star's spectrum to a curve of the proper temperature and know its distance, we can find its size — even though we may have no hope at all of ever measuring such a tiny angle directly.

10.2.2.2 *Particles to waves: the Bohr atom*

Our picture of the atom is now of a tiny, very heavy nucleus with a positive charge; around this in some fashion are arranged the negatively charged electrons. The obvious analogy to the Solar System, with the Sun in the center holding most of the mass and planets orbiting around, was seen right away (and still survives in oversimplified descriptions). If electrical force is substituted for gravitational force, for a single electron the classical mathematics looks just the same. Almost.

The biggest problem was that, according to classical electrodynamics, a charge going around in an orbit should radiate electromagnetic energy, thereby falling into a lower orbit. When the calculation was done, it was found that atoms should only last for a hundred-millionth of a second or so.

---

[7]The curve in Fig. 10.1 looks suspiciously like that for the Maxwellian distribution in Fig. 5.5. In fact they have a lot in common, but the mathematical shapes are subtly different. For our purposes the details are not important.

[8]There are two-photon processes in physics, but they're rare; it's hard to arrange things so that two photons arrive together.

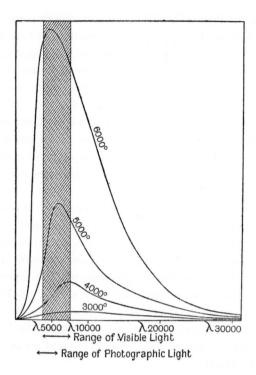

Fig. 10.1 Spectral curves of blackbody radiation of various temperatures. The height of a curve is the amount of radiation put out at that wavelength (per area of the blackbody surface), the latter measured in Ångstrom units, 10^{-10} m. Curves are labelled with the absolute temperature in Kelvin. The range of wavelengths visible to the eye is shaded. Notice two things about the curves: first, the maximum, where most of the energy is put out, moves from longer to shorter wavelength as the body gets hotter. At 3000° the maximum is outside the visible range, but some energy that the eye can see is still put out; more in the red than green or blue, so the object looks red. At 5000° the maximum is in the green-yellow, but enough is put out across the visible spectrum to make it look yellow-white. Above 6000° the object would look blue. In addition, the area under the curve, which measures the *total* amount of energy radiated, increases enormously with temperature. This diagram was taken from Jeans (1929); figure ©1929 Cambridge University Press, all rights reserved. Reprinted with the permission of Cambridge University Press.

(Another difficulty is that forces between the electrons in any atom, and there are several electrons in any atom more complex than hydrogen, mess up the picture completely.) The fact that atoms exist meant that classical physics was inadequate to describe them.

Well, if waves can be particles, can particles be waves? Louis de Broglie followed that speculation and found that, indeed, *particles have a wavelength.* It is normally very small, and gets smaller with increasing momentum, but on things the size of atoms an electron's wave nature becomes important. Neils Bohr worked it out this way: suppose electrons around atoms are restricted to orbits that are a whole number of wavelengths in circumference, so that the electron's crests always meet crests. If you give an electron in the lowest orbit just enough energy to move it up to the next, it will absorb the energy and move; if you give it not quite enough, nothing happens. So if you shine a continuous spectrum of light on a group of atoms, they will absorb only those wavelengths corresponding to possible energy transitions. Suddenly the absorption-line spectra of elements had an explanation! And after you shine the light, or maybe inject energy some other way, the atoms will seek to lose this energy by radiation, giving off an emission-line spectrum with lines in just the same places (because they correspond to the same changes in electron energy).

Going through the numbers, for hydrogen everything works out beautifully. Other atoms don't fit so well, or at all; and there's still the difficulty of the interaction of electrons among themselves in most atoms. Still, the fit convinced physicists that they were indeed on the right track. And even when the picture became more complicated with the New Quantum Theory, this idea of electrons making transitions among discrete energy-levels in the atom remained, and remains.

From the spectrum of an element one could construct the energy-level diagram of an atom, similar to Fig. 10.2. But it was found that there were some transitions that just didn't happen, for reasons that the Old Quantum Theory could not explain. The reason for these "selection rules" was one of the motivations for research leading to the New Quantum Theory.

Note that there is a highest bound energy in the diagram. If you give an electron more than this much energy you remove it from the atom completely, that is, you *ionize* it. (This *binding energy* of the outermost electron in an atom explains a great deal of the chemical behavior of the elements, since chemical reactions are rearrangments of the electrons shared by more than one nucleus.) For hydrogen this corresponds to a wavelength in the ultraviolet, not found on Earth because the atmosphere shields us but quite common in space. So electrically neutral atoms of hydrogen (and other elements) can be split up by the Sun's radiation into positive and negative *ions*, which will then feel any electric and magnetic forces that there happen to be. (This point seems to have escaped astronomers in the first half of the twentieth century, at least among our authors and their sources.)

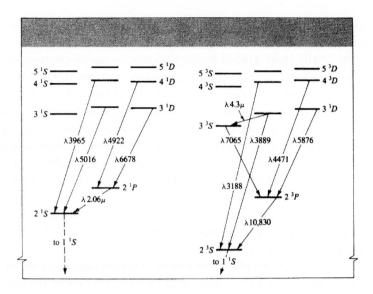

Fig. 10.2 An energy-level diagram for an atom with a single electron, in this case helium with one electron missing (which happens often enough in space). An electron in the gray area at the top has a positive energy and thus has escaped; the slots shown have negative energy (an electron in one of them is bound to the nucleus) and get lower in energy as you go down the page. Quantum-mechanical labels for the various states are shown; they were only figured out in the New Quantum Theory. The important transitions leading to strong spectral lines are shown as arrows, with the wavelength of the line given in Ångstroms (or in μ, ten thousand Ångstroms). Note that the wavelength of the line is longer as the levels are closer together (and thus the difference in energy is smaller). Note also that there are many transitions that don't lead to lines; these "forbidden" transitions were only explained by the New Quantum Theory. This diagram is taken from Osterbrock (1989); ©University Science Books, reprinted with permission.

10.2.3 The New Quantum Theory

The Old Quantum Theory presented some serious challenges to the thinking of scientists. How could something be, at the same time, a wave and a particle? As pointed out in chapter 5, a particle must go through one of two holes in a screen, while a wave goes through both. And yet when the experiment was set up, electrons produced interference just like light (the experiment is very difficult because the wavelengths involved are so tiny, but it has been done). When you set up a double-slit experiment, with light or electrons or whatever, and you try *at the same time* to detect which slit the electron or photon went through, the intereference pattern disappears: the detection disturbs the detectee. It turns out that you can't actually set up an experiment as a multiple-choice question: "You are a (a) wave, or

(b) particle." You can only ask, "Are you a wave?" or "Are you a particle?" (you'll get the answer "yes" for each).

The New Quantum Theory almost could have been produced to bend the minds of anyone who had gotten past the strangeness of the Old Quantum Theory. It is more abstract, more difficult and complicated mathematically, and presents more serious problems to anyone trying to produce a mental picture of just what is going on. However, it gave answers where the Old Quantum Theory gave none, and gave the *right* answers where the former theory was inaccurate. I won't pretend to give a full presentation of it; however interesting in itself, that would take us too far from our purpose, and require a great deal of time and space. I will just give a rough outline.

To every object in the universe is associated a *wave function*. This is a mathematical expression, generally complex (involving the imaginary numbers, those whose square is negative, so making it even more difficult to visualize). To predict what will happen when you make an observation of the object, you apply a mathematical *operator* to it of the proper type. To find out what the theory says you should get as the momentum of an electron in an atom, for instance, you apply the momentum operator to its wave function. You will get an expression that gives the *probability* of the various possible values of momentum; and that's the best you can do. Uncertainty seems to be a basic property of physics at this level (a feature which has led to a great deal of deep philosophizing, some of it worthwhile, much of it ill-informed). If you set up the situation properly you may get a nicely peaked probability, saying you're almost certain to get an observation very close to a certain value; and if you average over lots and lots of atoms you get something like a straight answer.

The Bohr orbits around the atomic nucleus were replaced by *orbitals*, less well-defined in location (they can only give the probability of an electron being in a certain place) but every bit as well-defined mathematically. And they give an energy-level diagram (there is an operator for energy) that looks very similar to Fig. 10.2, so the ideas of electrons making transitions among the levels is retained.

With this kind of apparatus, scientists could work out such things as the probability of a transition from one energy-level to another, thus giving a theoretical basis for the selection rules. And just as important, they could calculate what would happen under conditions that could not be reproduced in a laboratory, such as the extreme vacuum of space or the extreme temperatures of stars.

Two types of related advance in this area of physics came about at the same time, though they are not strictly part of the New Quantum Theory. The first is the experimental detection of the neutron, a particle about the mass of the proton but having no charge; it is found in the nuclei of atoms, making up a good part of the mass. It alone upset the idea of a two-particle universe and gave warning of the much more complicated zoo of particles yet to come. The other is the development of a set of rules about what you can actually do in a particle-physics reaction. Not only must you conserve energy and momentum, but also things like baryon number. Without going into details, this means that you *cannot* have an electron and a proton come together in a matter-annihilation reaction. You must balance a proton with an antiproton, or an electron with a positron (which by now had been detected); that is, only *antiparticles* can annihilate each other. Each particle has its antiparticle, one with the same mass (and other features) but the opposite charge and baryon number. If you can run antiparticles into each other, then you can convert all you mass into energy. (You can also go the other way, by running two photons of sufficient energy together to produce an electron-positron pair. This is much harder to do, and hadn't been done in a laboratory by the time any of the books we'll look at had been written.)

I don't mean to imply that everything had been worked out by mid-century. What happens *inside* the nucleus was still not terribly clear, many details of particle-physics and nuclear reactions had yet to be filled in, and most of the desirable calculations (for anything apart from the simplest situations) were simply impractical. There were useful answers, but also plenty of questions to keep physicists busy.

10.3   Observations of stars

Less spectacular (and mind-bending) than the new physical theories, but just as important for our purposes, was the mass of detailed and accurate observations of stars that had been accumulated by the second quarter of the twentieth century. Painstaking work with more precise instruments had yielded good distances to dozens of stars and less reliable results for many more. Detailed spectra of stars were common, at least to the point of allowing a general classification of each. And photography had allowed the accurate measuring of the brightness of stars (which Herschel, we recall, had felt as a major lack; see Sect. 6.4.1.2). Careful observation of double stars had allowed the calculation of some stellar masses. From this, a great deal can be deduced.

Distances and apparent brightness can be turned into intrinsic brightness, since we know how the intensity of light falls off as things get farther away. Fitting the spectrum to a black-body curve gives an effective temperature, which together with intrinsic brightness gives the size of the star. For a few stars, not very many yet but enough to allow a start, we now have mass, luminosity, temperature and size; we must now somehow build up a theory to match them.

Probably the most useful way to set out information on stars is the Hertzsprung-Russell diagram, shown in Fig. 10.3.[9] This plot uses several things which are astronomical idiosyncracies, so it needs some explanation.

Along the vertical axis is plotted luminosity, the intrinsic brightness of a star (with the effect of distance removed). Remember that astronomers use a scale of brightness inherited and refined from that of the ancient Greeks, in which a lower number is brighter. It's not too strange when one is talking about "stars of the first rank" or "second rank" as they were then; but when things were standardized it led to the strange feature of a star of magnitude -1.5 being brighter than one of magnitude -0.5, and both outshining one of magnitude 2. The Sun is among the fainter stars on this plot, there at about magnitude $+5$. The scale is logarithmic; a star 100 *times* brighter than another has a magnitude 5 *less* than the other. For one magnitude difference, the ratio is about 2.5.

Along the horizontal axis is plotted spectral type. This is determined by a careful analysis of the spectrum, comparing the strengths of the various absorption lines of elements, and is given by a letter and number; the Sun is a G2 star. It correlates with temperature, so that a B star is hotter than an A star, which is hotter than an F star, and so on. It also correlates with color, as one would expect for objects that are roughly blackbodies, with B stars being blue, G yellow, and M stars red. Because spectral type only appears in discrete steps, stars being G2 and G3 but not G2.71 for

---

[9]I am being intensely conservative in using the word "probably" here. All astronomers who have anything to do with stars seem to think instinctively in terms of the "HR diagram" and its variants. Stellar astronomers, of course; anyone building galaxies, since observationally they're made mostly of stars; anyone *using* stars to deduce things about galaxies; those who investigate the tenuous gas and dust between the stars; even cosmologists have been known to show an HR diagram somewhere in their talks. I can't think of anything in astronomy, even the Hubble plot, which has been so popular, or so useful. (The Hubble plot has to do with cosmology. Unfortunately, we are not going far enough into that subject to need to look at one.)

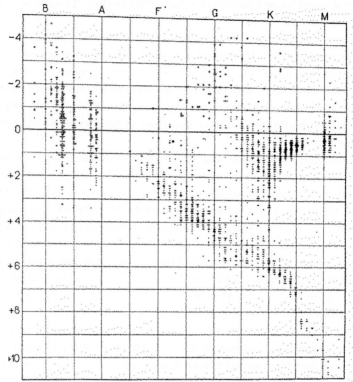

Fig. 3. Statistics of Absolute Magnitude and Spectral Type.

Fig. 10.3 The Hertzsprung-Russell diagram, from Eddington (1926). Along the horizontal axis is spectral type, which correlates to temperature (higher temperatures to the left, a strange way to plot something but not the only idiosyncrasy astronomers have) and color (bluer to the left, redder to the right). Along the vertical axis is the intrinsic brightness of the star, in absolute magnitude (arithmetically smaller numbers are brighter). The Main Sequence extends from the upper left to lower right; the Red Giant Branch is at the upper right. The Sun is on the Main Sequence at about magnitude +5, spectral type G. The vertical lines of dots representing stars are artifacts of the way the plot is set up; if temperature is used as a horizontal axis instead of spectral type, the Main Sequence is a continuous band. Figure ©1926, 1988 Cambridge University Press, all rights reserved. Reprinted with the permission of Cambridge University Press.

instance, stars plot in vertical columns here. If we used temperature or color the columns would disappear.

And it is often useful to plot color directly, rather than working out spectral types. (In fact it's much easier to collect the data: you take pictures in two color filters and subtract the star's magnitude in one from its magnitude in the other, rather than taking spectra and doing a detailed analysis of it.) Then the plot is, strictly speaking, a *color-magnitude diagram* and not an HR diagram; but for our purposes they do the same thing.

Sir Arthur Stanley Eddington, from whose book I took this figure, gives two cautions about this particular one. Both stem from the fact that there are far more faint stars than bright stars in any particular bit of space. First, it means that the very bright stars (at the top of the diagram) are found rarely, so are mostly far away, where the distances are more uncertain. This turns into an greater uncertainty in brightness. You might think of this as a kind of vertical smearing, more serious as you go up the figure. Second, the lower part of the figure should have far more stars in it, if you want to compare the number of stars in any bit of space. In this figure, the Sun, at magnitude +5, looks like an unusually faint one. In reality, there are many more stars fainter than the Sun than there are brighter ones.

The first of the major features of the HR diagram is the Main Sequence (called by Jeans as well as Eddington the "Main Series;" I do not know when the terminology changed), extending from upper left to lower right. (Remember the vertical smearing at the top: the Main Sequence in B and A stars could actually be as thin and well-defined as it is for fainter stars.) The Sun and most stars fall on this sequence. The second feature is the Red Giant Branch, up there at the top right. K and M stars especially fall into two distinct types: bright giants, and faint dwarf stars. (The astronomers who set up the terminology did not designate anything as "normal-sized.") Red Giants are in fact very big, as calculated by the black-body formula and in a few cases as measured by an interferometer. To the upper left of the Red Giant Branch, the sprinkle of very bright F and G stars are mostly or entirely pulsating variables. Way down on the lower left corner of the diagram, not appearing in this particular one, would be the enigmatic White Dwarf stars. Only a few are known because they're so faint.

Later on we shall see another plot, luminosity against mass, that summarizes the observational material; and one can plot on top of this HR diagram the sizes of the stars. But what we have in Fig. 10.3 is essentially the first set of data that any theory of stellar structure and evolution must match. It must explain why most stars fall along the Main Sequence, and why there is a great gap below it; why there is another clump on the Red Giant Branch, and why the stars to the upper left of it are unstable to pulsations; and in general why there are stars in some places on the diagram, and not others.

10.4 The Milky Way and the nature of the nebulae

Two related problems of long standing have been all but solved by the time the next chapter opens, so that areas we have spent much time on will almost escape mention.

The first is the size and shape of the "stellar universe," the Milky Way. By the 1930s it became clear that there exists something within this structure that absorbs light, so that distant stars are dimmed by even more than distance. Our next chapters take place before then, but observations and interpretations were already tending in that direction. In some places (like the Coal Sack) there is a greater concentration of this stuff, so it's easier to see, but it also appears generally in the plane of the Galaxy. Crucially, this absorbing stuff makes the light coming through it redder (exactly the way a sunset has red colors), so it can be traced by something more than just the dimming of starlight. This realization, plus other clues from various sources, showed the Sun to be well away from the center of the whole structure (though still very close to the central plane) and gave dimensions that are not too far from those accepted today.

The absorbing medium, interstellar dust, means we cannot see very far in the plane of the Galaxy, and of course cannot see beyond it. That conveniently explains why the nebulae seem to avoid the Milky Way, something noticed long ago (and drawn in Newcomb's picture), if they are in fact objects outside it. As we open the next chapter it is becoming accepted that nebulae are probably external galaxies, something like the size of the Milky Way and very, very distant.

Chapter 11

Sir James Jeans, *The Universe Around Us*, First Edition, 1929

The year is 1929 and everything has changed. The Great War, of course, has swept away empires, kingdoms and some basic certainties of the nineteenth century along with many of its inhabitants. In science, revolution is also in the air. Newton's great achievement has been found to be only approximate, and the strange ideas of the Quantum Theory and Relativity are being worked out. They show great promise in explaining much that was intensely puzzling to the past century as well as some niggling details that hadn't attracted much attention. However, they are still not well understood, and physics as a whole has no idea how to work with them in many situations. Among these are some areas of astronomy.

James Jeans began by working primarily in physical theory, on the theory of gases and the implications of the quantum theory as it began to be developed with the new century. He did much original work and much that has stood the test of time: in current astronomical papers one still finds references to the Jeans Mass (which we will look at in some detail), and the Jeans Equations of Stellar Dynamics form the basis for much work in the study of galaxies. In the first third of the twentieth century he is clearly an important scientist.

As a mature scientist he increasingly turned his hand to popular science, both in the new medium of radio and in a series of books for popular readers. I have chosen *The Universe Around Us* (Jeans, 1929) as the most comparable to previous works I've looked at in this study in subject and coverage, out of many possible choices. It also has the advantage that Jeans published, nearly at the same time, a book with almost identical subject matter but with all the mathematics left in: *Astronomy and Cosmogony* (Jeans, 1928). It will be useful to refer to the latter work in a couple of places, though of course it is well beyond the reach of the layman.

11.1　Audience, aim and content

In his short Preface Jeans characterizes his book as

> a brief account, written in simple language, of the methods and results of
> modern astronomical research, both observational and theoretical. Special
> attention has been given to problems of cosmogony and evolution, and to the
> general structure of the universe. My ideal, perhaps never wholly attainable,
> has been the making of the entire book intelligible to readers with no special
> scientific knowledge.

Although he quotes numbers on occasion, he does not ask the reader
to do any mathematics and his algebraic formulae are carefully confined to
footnotes (for example, pp. 95-6, 198). He does expect that his readers know
how to square a number (p. 95), but this is about the limit of mathematical
skill he requires of them. His readers, then, are not Sir John Herschel's
numerate elite of the public; but the complexity of the material he has
to cover means that a good degree of attention and reasoning ability are
necessary to follow his explanations.

Part of his material has been given before in wireless talks and University lectures (according to the Preface), but all has been completely
rewritten, so we need expect none of the style or limitations of verbal communication (and, in fact, I could find none).

Note, however, that he is *not* presenting a balanced survey of the whole
of the science of astronomy. He is concentrating on the results of *modern*
astronomy. This is very important in any overall evaluation of the book,
and we will come back to this point.

In a slightly different formulation, he explains that he is "setting out
to explain the approximation to the truth provided by twentieth-century
astronomy. No doubt it is not the final truth ..." (p. 8), and expands the
latter thought in a jigsaw-puzzle similie a few pages later (p. 15). On the
other hand, "modern science, eschewing guessing severely, confines itself,
except on very rare occasions, to ascertained facts and the inferences which,
so far as can be seen, follow unequivocally from them" (p. 9). So at the
outset Jeans is claiming to have a high level of confidence in his results —
they are not mere guesses, though they are not to be considered complete.
This is another matter to which we will return later.

The contents of the book are divided into seven chapters in this way:

Title and content	Pages
Introduction	15
[historical background and overview]	
Exploring the Sky	73
[observations, plus some relativistic cosmology]	
Exploring the Atom	58
[the Old Quantum Theory]	
Exploring in Time	45
[calculations of ages]	
Carving out the Universe	55
[formation of galaxies, stars and planets]	
Stars	70
[stellar formation, structure and evolution]	
Beginnings and Endings	28
[extrapolations to the distant past and future]	

We note that Jeans has been as good as his word. All the subject-matter of Herschel's and Airy's books, and almost all of Newcomb's and Ball's is covered in about forty pages of the second chapter. Indeed, if we leave out the very speculative cosmology of Newcomb and Ball we can say that Jeans has fit all of nineteenth-century astronomy into about ten percent of his book. Geodesy, for instance, a full chapter in Herschel, receives a single paragraph. Jeans is indeed concentrating on *current* results. It is not much of an exaggeration to say that the questions he spends almost all his time on could not even have been formulated by the earlier authors.

11.1.1 *The Jeans picture*

In brief, Jeans is concerned with the formation, structure, and evolution of galaxies, stars and planets. This is what he means by "cosmogony," a word that has gone out of use.

There are two physical processes which underlie Jeans' picture. One is the loss of mass (of stars and galaxies, particularly) due to radiation. According to the combination of quantum and relativity theory, mass and energy may be converted into each other; so light (and all radiation) carries away a certain, calculable, amount of the mass of a radiating object. Because the Sun shines, it weighs less today than yesterday, and will weigh less tomorrow. The difference is slight over days, even over periods of millions or billions of years. But if we have to deal with stars lasting ten or a hundred or a thousand times a billion years, the loss of mass by radiation can be substantial.

The second process is the fragmentation of an object by "gravitational instability," in which parts collapse to form smaller objects. Consider a big, uniform mass of gas: that is, something larger than anything we need to look at,[1] and the same temperature, pressure and density all the way through. We also assume it to be in equilibrium: all the forces on it, particularly gravity and gas pressure, exactly balance out.

Now we disturb it slightly, as in any calculation of stability. (There's no need to worry about what and how; the universe is always a disturbing place.) That means that there will be sections denser than before, and other parts rarer. In the denser parts, the gas pressure will go up and if that is the only effect, it will push the dense part back toward equilibrium again. In fact it oscillates a bit: this is a sound wave. If we think about gravity (as we almost always do in astronomy), for anything in normal experience it's a tiny effect: the force of gravity exerted by the air in a room at normal temperatures, whether it's made slightly denser or not, is entirely negligible compared to gas pressure.

But if the denser bit is big enough, its own gravity will become important. Pressure, and small changes in pressure, will stay the same no matter how big a chunk of gas we take; but gravity grows with mass. At some particular size a disturbed bit of gas will have enough gravitational force to overcome the increased pressure, and will get even denser. In a runaway situation, we have gravitational collapse. The mass required is larger if the gas is at a higher temperature and lower if it's at a higher density, and depends slightly on what the gas is made of; but given all that it's calculable, and given the name "Jeans Mass" for that situation.

We started with a slightly disturbed mass of gas. Unless things are very exactly balanced, there will be at least a tiny bit more motion in one direction around the center than in another. That motion will get larger as the mass gets smaller: angular momentum is conserved. So the collapsing mass will not contract forever, but find a new equilibrium in which some of the gravitational force is balanced by centrifugal force. Depending on how much angular momentum the object has it will be a more or less flattened sphere, an *ellipsoidal figure of equilibrium*. We will hear more about these.

---

[1]One could think of an *infinite* mass of gas, a whole universe filled with it. Apart from the trouble many people have with the very idea of infinity (not a trivial matter!), Newtonian physics can be unreliable if things are truly infinite. Instead, then, we'll just say that this mass extends to a large distance, an only look at what happens to it locally.

Now for the Jeans picture.

In the beginning was the "primaeval nebula," a vast and very diffuse mass of almost homogeneous gas. Under the mechanism of gravitational instability parts of it began to contract, growing denser and spinning up (as tiny amounts of angular momentum are brought in to smaller regions). These became galaxies. What we see as the centers and bulges of galaxies are made up of hot gas which has not yet formed stars, and is too hot to do so yet. Over time the galaxy will lose mass by radiation and shrink, leaving behind chunks of gas at its edges which in turn collapse by gravitational instability into stars. The shapes assumed by galaxies as they evolve are ellipsoidal figures of equilibrium, the theory of which is set out in *Astronomy and Cosmogony* and to which Jeans has made significant contributions.

Stars themselves are not gaseous, at least not in their centers (and for much of the book Jeans treats them as completely non-gaseous). Instead they act like liquids, which cannot be compressed but do change shape under forces. They are powered by electrons in their atoms combining with protons and so converting the entire mass of both to energy, in a reaction that does not depend on temperature or pressure (until temperatures get so high, as they are in the visible gas of galaxy bulges, that they cannot combine at all). Close binary stars are formed when liquid stars shrink (by radiating away mass), leaving behind rings of matter at their edges that contract under gravity to form another star; or very, very rarely when two stars pass close enough to each other to raise a great plume of matter by tidal forces which then collapses into another star. Stars live for some hundred billion years or so, eventually radiating away all their mass. All the stars in the Milky Way were formed at essentially the same time.

The Solar System was created when another star passed very near the Sun, within a very few radii, and pulled out a plume of the Sun's material by tidal forces. This plume fragmented, collapsed and cooled into the planets we see today. The satellites of the planets were formed in the same way, when the new planets on very eccentric orbits passed close enough to the Sun to have material drawn out of them in turn. A former planet passed too close to Jupiter and was fragmented into the asteroids.

11.2   Hindsight and Sir James Jeans

It is hard to know just how to approach such a mixture of penetrating insight and sheer error. Where Jeans has a good large idea, the details tend to be completely mistaken, and vice versa. Gravitational instability is

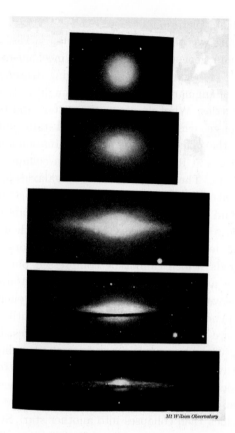

Mt Wilson Observatory

Fig. 11.1   A series of nebulae in the order required by Jeans' theory of galaxy evolution, their shapes resembling ellipsoidal figures of equilibrium. He pictures NGC 3379, at top, as early in the process, large and not rotating enough to affect its shape. NGC 4621, next, has shrunk enough from radiating away mass that its angular momentum makes it somewhat flattened. The flattening continues with NGC 3115, in the center. By NGC 4594, second from bottom, the edges of the galaxy have started to break up into stars or star clusters. NGC 4565, at the bottom, is mostly made up of stars (the diffuse glow in each galaxy is taken to be gas not yet formed into stars). From Jeans (1929); figure courtesy of The Observatories of the Carnegie Institution of Washington.

the basic mechanism for forming galaxies and stars, but his picture of the structure and evolution of a single galaxy is quite incorrect. Galaxy bulges are masses of unresolved stars, and stars can form anywhere in the galaxy (though they tend to do it within the disk). Stars are powered by nuclear reactions, but protons do not combine with electrons at all, much less in a total-conversion way. Stars are gaseous, not liquid (except for relatively

Fig. 11.2   A photograph of the center of the nearby galaxy M31. We see it through a screen of stars in our own Milky Way. According to the Jeans picture, the part of M31 visible here is gaseous, without stars, and too hot to form stars or participate in the matter-annihilation reaction that powers stars. From Jeans (1929); figure courtesy of The Observatories of the Carnegie Institution of Washington.

rare examples of extreme density, where the state of matter is quantum-mechanically exotic). No stars form by the instabilities of a shrinking liquid mass in rotation. Stars do radiate away part of their mass; but since they live at most a few billion years, this does not have an important effect on their evolution.

As far as getting things *right*, it would probably be best to leave Jeans aside and start over. For reasons I'll get to toward the end of the chapter I believe that would be a mistake. In the interim, however, we have the problem of analyzing Jean's mistaken results from the standpoint of the layman reading his book. Because the ideas that turned out to be right are so mixed up with those whose fate was less happy, I will not try to present a summary of the specific errors in material. Instead, I will organize my analysis around the various techniques that Jeans uses to present his conclusions.

Fig. 11.3   The outer regions of M31, showing it (according to the Jeans picture) breaking up into stars or star clusters. From Jeans (1929); figure courtesy of The Observatories of the Carnegie Institution of Washington.

11.2.1   *Jeans and theory*

11.2.1.1   *The stability of gaseous stars*

I must beg my readers' indulgence for spending time on this particular calculation, though it is rather esoteric and complicated. I have had the occasion to go through Jeans' mathematics in detail (found in Jeans (1928), pp. 117-128, §105-115) and explore a bit further on my own (with results published in Whiting (2007)), so it is now very familiar to me; and it brings up some subtle but important points.

Jeans builds a *model* of a gaseous star. That is, he writes down the differential equation of motion, adding up forces (gravity, gas pressure) and setting them equal to the change in momentum of a piece of the star; and the differential equation of energy, saying that energy generated in a bit of the star can raise that bit's temperature, expand it against pressure, or flow elsewhere. In order to make the result managable at all he must make a number of simplifying assumptions, such as that the material is same throughout and that the star is perfectly spherical.

Now he inserts some assumptions (good ones, as it turns out) about the interaction of radiation and matter (the *opacity*) and other details. And he puts in an energy generation rate that depends on the temperature and density to some power.

To see that this is reasonable, consider a chemical reaction between molecules of a gas. The reaction occurs when one molecule hits another; it can't happen between molecules that are separated. If you double the density, you (i) double the number of molecules in a certain region available to hit others, and (ii) double the number of molecules there are to be hit. So in a simple two-molecule reaction, doubling the density multiplies the number of collisions by four. The reaction rate (all else being equal) goes as the *square* of the density. More complicated reactions will give different specific numbers, but a reaction rate like this is not a bad first guess at anything.

When the molecules hit, they must somehow rearrange; that's what a reaction means. Some of the energy of collision must go to breaking a part of the molecule's structure. If the temperature is higher, there are more molecules out of the total that have enough energy to do the initial rearranging, and thus to do the reaction. (Remember the Maxwellian distribution, Fig. 5.5, and the comments I made on it in Chapter 5.) The details are complicated, but over some restricted range of temperature at least, the rate of reaction (once the molecules hit each other) can be approximated as a power of temperature.

If the (completely unknown) nuclear reactions that power a star behave like chemical reactions, this is the kind of effect we expect temperature and density to have on them. It is certainly a reasonable place to start.

Last, Jeans imposed equilibrium. That is, the material in a bit of the star (since spherical symmetry has been imposed, this is a shell at a given distance from the center) is not moving, and as much energy flows out of a shell as is generated and flows in. This completes the model.

Now Jeans perturbs the model: he imposes a small disturbance, and calculates whether it grows or shrinks. Here again he must simplify. There are an infinite number of ways in which even a spherically symmetric star can move: the center can expand a lot and the edge not at all, or vice versa; the temperature can rise in the center and sink at the surface, or vice versa; and everything in between. Jeans assumes a uniform expansion of the star and a uniform heating or cooling of the temperature.

Jeans found that, for any even moderate increase of energy generation with density and (more importantly) temperature, his star would go into a growing pulsation. The size would alternately shrink and expand, getting farther and farther from equilibrium each time. It was unstable.

Next he considers an idea put forward by H. N. Russell, in which the energy-generating reaction (whatever it is) does not happen at all below a certain temperature, and at a steady rate above it. *Without calculation* he asserts that the instability would be more violent, and thus that Russell's model is ruled out. In fact Russell's is the only gaseous model he mentions in *The Universe Around Us*, saying, "Mathematical analysis then shews that ... [Russell's star] would be in the state of a keg of gunpowder with a spark at its centre ..." (p. 287).

What appears to be happening physically is this: when the star is compressed the energy generation rate increases, which causes it to heat up and expand. But the inertia of outflowing stellar material carries it beyond the size at equilibrium; the energy rate drops below that of equilibrium (since it is now less dense, as well as cooler) and it contracts again. Over the cycle the extra energy generated during the contracted phase is greater than that lost during the expanded phase, and the pulsations grow. If this goes on long enough, the star blows itself apart.

By this calculation a gaseous star with a strongly temperature-dependent energy source is ruled out. Jeans proceeds to build liquid stars powered by an energy source that is not affected by temperature at all (at least in the stellar environment), although strictly speaking he has only ruled out the combination of gas-law behavior and temperature dependence and need only drop one of these features.

In fact, even taking the calculation as it stands Jeans has not, strictly speaking, ruled out either. He has ruled out a simplified version of a star perturbed in a certain way, and has shown that an *infinitesimal* perturbation will grow. Once the pulsation becomes big enough, the assumption of a small disturbance is no longer good and the calculation no longer applies. And the assertion that Russell's reaction-rate law would be even more unstable is an assumption (and, I think, not a good one), not a calculation of even this level. Here we have an example of an overinterpreted calculation; Jeans has not proved what he says he's proved.

The calculation, even within the simplifications, is flawed. This was concluded not by someone finding a mistake in Jeans' algebra, or pointing out an error in his physical mathematics. Several years later T. G. Cowling attacked the problem in an entirely different way and concluded that gaseous stars powered by temperature-dependent reactions were indeed stable (Cowling, 1934, 1935). He thought that Jeans' mistake lay in the form he assumed for the pulsation-perturbation, in that it wasn't a normal mode (see Sect. 2.6). My own investigation, many decades later, led me to believe that the form of the temperature-perturbation is the more important assumption. But I cannot *prove* it, because the situation is too complicated to treat with Jeans' methods if the assumption is relaxed. Just reproducing his calculation (to check his algebra and the effect of various assumptions) requires many pages of mathematics; if one tries to allow for more a more complicated situation the calcuation becomes truly horrible, beyond anything practical. Nowadays we would feed it into a computer, but in 1929 that was not an option.

A situation I think is similar, but simpler, is that of a pot of water on a stove. If it is heated from below eventually the water forms convection cells, in which hot water from the bottom rise to the top; cooling there, the water sinks to the bottom around the edges of the cell. If, however, one calculates stability on one dimension (which does not allow water to rise in one place and sink in another) the pot comes out to be unstable: it blows up. By simplfying the situation too much, one gets the wrong answer. This is what I think Jeans has done.

It no doubt sounds a bit strange for me to assert that Jeans' calculation was too simple, after spending several paragraphs describing it (actually doing the mathematics is much more complicated). Certainly complicated mathematics is nothing new in astronomy; there are hundreds of pages of equations, for instance, in the *Mécanique Céleste*, the basis of much of nineteenth-century astronomy. But recall that Laplace only dealt with gravity, and much of time only with point-masses. Jeans, even to set up his simplified equations, had to deal with not only gravity but the temperature-pressure-density behavior of gases; the interaction of radiation with matter; flows of energy of various kinds; details of very imperfectly-understood nuclear physics; and had to make some judgement about what was important, and what could be simplified away. This is my main point: astronomy had become so complicated by 1929 that even a very simple situation was difficult to analyze, and perhaps impossible to do rigorously with the tools at hand.

11.2.1.2 *Relaxation and the equipartition of energy*

We come now to what is, in *The Universe Around Us*, the starting-point of Jeans' picture.

Recall the idea of relaxation, of a system of particles or bodies disturbed in some way, and then allowed to interact so as to find a steady state. In Chapter 5 I gave the example of molecules eventually finding their way to a Maxwellian distribution of speeds. If you refer again to Fig. 5.5, you'll note that molecules of different masses relax to different distributions at the same temperature. If you were to mix them, the heavier molecules would be moving more slowly, and at such a rate that their average kinetic energy would be the same for each kind. This is an example of the equipartition of energy.

Stars are not molecules and gravity is not the same as the forces involved in the collision of molecules. However, a swarm of stars allowed to relax by interactions among themselves will also tend toward a Maxwellian distribution, at least in any limited region, one in which there is this equipartition of energy among the various kinds of stars.

If given details about the interaction one may calculate a sort of time-scale for how long this approach will take. For gas molecules at room temperature it's a fraction of a second, so that if you mix a sample of nitrogen held at 0° with one at 100° you have a uniform sample at a temperature in between within a fraction of a second.

With stars as far apart as they are in the Milky Way, and moving at the average speed they are moving, the time-scale for evening out a deviation from a Maxwellian distribution (achieving equipartition of energy) is very long, well into the billions of years. Jeans quotes a result by an observer that shows stars pretty close to equipartition (p. 161); and asserts, "The age of the stars is, then, simply the length of time needed for gravitational forces to bring about as good as aproximation to equipartition of energy as is observed," (p. 162) and goes on to give a figure of "from 5 to 10 millions of millions of years" (p. 164). This span of time, roughly a thousand times as long as we now use for stellar ages, conditions the whole of Jean's picture. Everything else is chosen to fit.

Setting aside any question about the details of the calculation (which is found in Jeans (1928), and no less complex than most things there), as well as with the sparseness of the observational material, there is one overwhelming problem with this idea. It appears on the next page (dealing with modifications of the orbits of binary stars): "It is impossible to discuss how far they have travelled along the road to equipartition without knowing the point, or points, from which they started" (p. 165). That is, if stars are *formed* more or less in equipartition, it will take no time to get there; if they are formed with a much different distribution of kinetic energies, it will take them a long time to relax to a Maxwellian form.

In *The Universe Around Us* Jeans does not specify where his time-calculation starts. In his picture of galaxy and star formation there is no detail concerning how new stars might have their velocities arranged; in fact he refers to "the hypothetical primaeval chaos, about which, from the nature of the case, observation cannot have anything to say" (p. 227). In *Astronomy and Cosmogony*, pp. 320-2, §288-9, he bases his calculation on how the dynamics of stars changes as they lose mass by radiation. In other words, he *assumes* at one remove the kind of time-span he is looking for.

It's hard to overemphasize the effect of this circular calculation. By adopting a time-scale a thousand times longer than we now use, he places his stars into the situation of radiating away a significant fraction of their mass, and of needing something like the total conversion of mass to energy to live that long, along with the other details of his picture. But his determination of the time-scale depends on the assumption of stars radiating away a large fraction of their mass.

This is not, of course, a contradiction or a logical fallacy. A change of mass by radiation inducing a change in stellar dynamics requiring a long time to adjust to equipartition all together form a consistent picture. But it is a circular sort of logic, and we will see more of this.

11.2.1.3 *Atomic theory*

In this heyday of the Old Quantum Theory there was quite a bit of elbow-room to speculate in. The new ideas had opened up possibilities beyond the imagination of classical mechanics and had not yet become definite enough to exclude many possibilities.

The basis of Jeans' stellar power source was a hypothesized reaction in which one of the orbital electrons of an atom combined with a proton in the nucleus, and thereby converted all the mass of both particles into energy. The principle of conversion of mass to energy had been established, though details were yet lacking for most ways of actually doing it, and this seemed a possible route.

Note that this was a *hypothesized* reaction, one that had never been seen and for which there was nothing in the way of direct (or even any specific indirect) evidence; the best one could say is that it had not yet been ruled out. In fact it was known not to happen in atoms on Earth, as classical mechanics called for (p. 132), so on top of the hypothesis of its existence one had also to assume that (for reasons unknown) it happened only within stars. Next Jeans assumed that it was not affected at all by temperature or density (a step that was not strictly required by his gaseous-star stability calculation, once he adopted liquid stars, but he made the assumption anyway), except that under extreme conditions of temperature (when atoms are completely ionized) it stops completely (pp. 325-6). By now the characteristics of the required way of generating energy had led to Jeans concluding, "it becomes clear that stellar radiation cannot originate in types of matter known to us on Earth" (p. 304).

The energy given off by this type of matter, made up possibly of su-perheavy atoms (p. 305), was assumed to follow an exponential-decay law (with a half-life), like the radioactivity of uranium or radium on Earth. A certain fraction of the reactable material in a given type of atom would react over a given time; a star was born with a mixture of fast-reacting atoms and slower ones (pp. 308-9). That leads to a gradual fall-off of en-ergy production as the star aged and lost mass through radiation.

What a marvellous stack of assumptions! Jeans' attitude, of course, was that he had been forced into each by observational material, and there had been no other choice, that this theory had grown directly from the facts. It is more accurate, though perhaps not totally fair, to see them as Jeans assuming his way out of any difficulty that happened to arise. From the standpoint of establishing how reliable this extended theory might be, the best that can be said is that, from what was known about nuclear reactions at the time, none of these assumptions was ruled out.

The contrast between Jeans in 1929 and Sir John Herschel in 1869 is remarkable. Herschel, as we saw, was sceptical of thermodynamics and had serious reservations about how spectra worked. Jeans, on the other hand, plunged into the new theories, mastered what was known, and pushed them as far as he could — which turned out to be a bit too far.

What, then, can we say about Sir James Jeans and theory in summary? He was certainly capable of the most advanced and complicated calcula-tions, and had mastered the physical basis of astronomy as far as anyone at the time. The science had become so complex, however, that small and obscure mistakes and assumptions could go undetected and have serious consequences. In addition, the state of physics was such that there were many new ideas and no stable and rigorous structure to test them with. The result was that Jeans was too able to assume his way out of any difficulty, and to mistake a particular result for a general one.

11.2.2   *Jeans and observations*

Jeans was a theorist, not an observer. Since the days of Herschel and even Newcomb the divide between telescope operators and those who spent their time with esoteric physics had been getting ever larger. Yet in any kind of astronomy one must eventually deal with the observations themselves, and it is illuminating to see how Jeans performed in that area.

11.2.2.1   *Agreement: exact, rough and very rough*

Jeans is fond of the word "exact". After outlining his theory of the formation of galaxies, he says that the result should be a series of flattened bodies rotating at different rates, "and this is exactly what is observed" (p. 202; see Fig. 11.1). His wording seems to refer only to the fact that nebulae are rotating, and at different rates, though it's fairly clear that he means that their shapes are those of ellipsoidal figures of equilibrium.

In fact he is quite ready to take even rather rough agreement as support. The diagram of the planets of the Solar System, for instance (p. 235; see Fig. 11.4), is rather misleading. It shows a smooth, cigar-shaped envelope representing the tidal tail pulled off the Sun by the (hypothesized) passing star. But consider that Jupiter has more mass than all the other planets together, and Saturn in turn larger than all the remaining ones. An envelope that accurately followed the planets' real sizes would be far less smooth, essentially an egg-shaped blob around the two biggest. Similarly, Jeans claims that the number and relative size of the satellites of the planets is "exactly in accordance with the prediction of the tidal theory" (p. 238). First, the number of satellites should follow the smooth outline of the cigar-shape; this is only true if the tiny rocks that circle Mars (30 and 15 kilometers at their maximum length) are considered the equivalent of the Moon (1738 km in diameter) and the larger satellites of Jupiter (roughly 3600 to 5300km diameter). The fact that Mars is not intermediate in size between the Earth and Jupiter, as called for by the cigar, is explained by assuming it to have been more gaseous than Earth at birth, and thus to have lost more of its mass before it solidified; an idea that is supported by the fact that Earth and Neptune (marking the ends of the gaseous-at-birth planets) have "comparatively big" satellites (p. 238). In the latter statement the ratio of lunar mass to terrestrial (1/81) is held to be equivalent to that of Triton to Neptune (1/4760). It seems to me characteristic that Jeans says the size of the orbit of the Earth around the Sun "increases at the rate of about a metre (39.37 inches) a century" (p. 224), equating a

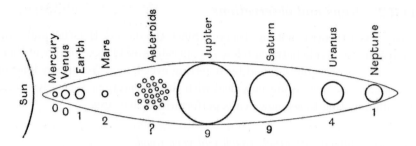

Fig. 11.4   Jeans' attempt to fit the known sizes of the planets into the theoretical cigar-shaped filament drawn off the Sun by a passing star, in accordance with his theory of their formation. Underneath each planet is a count of its satellites, another feature of his theory. In reality the total mass of the asteroids is less than that of the smallest planet, so the fit is not very good. From Jeans (1929); figure ©1929 Cambridge University Press, all rights reserved. Reprinted with the permission of Cambridge University Press.

rough measurement with one quoted to four figures: confusion of, or at least inattention to, vastly different levels of accuracy.

Elsewhere the agreement is even less exact. According to Jeans' theory of stellar formation and evolution,

> ... about one star in 100,000 ... may be attended by a retinue of planets. Another fraction, still small, although far greater than the foregoing, appears to have broken up as the result of excessive rotation,and formed binary or perchance multiple systems. But the destiny of the majority of stars is to pursue their paths solitary through space ... (p. 246)

However, when the number of multiple and single stars in the local neighborhood (as given by Jeans over pp. 256-61) is added up, 9 out of 25 are in multiple systems: 36%. This must be a lower limit, since there may be more stellar companions that haven't been seen. In any astronomical context, it is hard to consider a fraction of over one-third as "small." And this is from data given by Jeans himself!

11.2.2.2   *False and falsifiable*

And then there are things which are just not true. "The asteroids occur as a single swarm" (p. 242); well, no. Most have orbits between those of Mars and Jupiter, but they are well spread out along those orbits, and there are a few (the Trojans) at the same distance from the Sun as Jupiter. "As things are, the whole swarm can be explained quite simply as the broken fragments of a primeval planet," (p. 242) torn apart by Jupiter's gravity. (p. 245)

Again, no. Simon Newcomb had already demonstrated that the orbits of the asteroids were such that they could not have had a common origin (Newcomb (1878), p. 327), certainly within the last few million years. Beyond that the calculation was too subject to small uncertainties to make a categorical statement. Strictly speaking, it was possible at the time to argue that the asteroids had been one body long before, since there were at least a few billion years to play with. But such a statement can hardly be characterized as support for Jeans, and would require at least a mention of Newcomb's result. Jeans seems to have been simply unaware of it; and since it had been well-known for half a century, that in turn implies that Jeans was simply unconcerned about anything other than a superficial glance at observations.

There are observations easy to make, indeed already existing, but to which Jeans does not refer. In his picture of galactic evolution,

> The temperature at the centres of the spiral nebulae may be, and in all probability are, so high that atoms are stripped bare of electrons and so shielded from annihilation. (p. 325-6)

To check the temperature of the center of a nebula all one has to do is take a spectrum and fit a blackbody curve. Spectra of this type had been available for decades (One of M31, taken in 1912, appears in Russell et al. (1945)); in fact Jeans refers to spectra of M31 when he discusses the rotation of nebulae (p. 202). They look rather like stellar spectra of a type a bit cooler than the Sun, not at all like a very hot gas. Indeed, gas of the temperature Jeans requires would be extremely blue, noticeable in the crudest of spectra or on pictures taken with color filters; instead, it is rather red. Jeans cannot have been unaware that the observations existed to check his theory; he seems, however, to have been uninterested in them.

Later on in his investigation of the limits of stellar mass and luminosity, he dismisses the possibility of any significant population of dark stars:

> If a star ceased to shine, its gravitational pull would still betray its existence. Although we could not detect a single dark star in this way, we could detect a multitude. If two stars out of three were dark, we should probably suspect the existence of the dark stars from their effects on the motions of the remainder, so that general gravitational considerations preclude the possibility of there being a great number of dark stars. (p. 264)

This assertion of some kind of observation ruling out a dark population does not fit well with Jeans' realization in *Astronomy and Cosmogony* (p. 9) that beyond the very, very nearest stars we simply cannot see the faintest ones, and so we do not know how many there are. Here he appears to be *assuming* an observation, of a very vague kind, to support his statement. (That the observation he refers to was never made is shown by the fact that we now know most of the mass in the Milky Way to be made up of something invisible — exactly the situation he rules out.)

In another place, he assumes some suspension of known physics (the gravitational redshift) in order to fit observations of very hot stars into his picture (p. 313-4).

Jeans' skill with, and attitude toward, observations is thrown into high contrast when compared with that of another theorist we have looked at, Simon Newcomb. The latter, as we have seen, was critical and thorough in his evaluation of observations, probably the most reliable interpreter of them among all our astronomers. It is hard to escape the conclusion that Jeans, by contrast, was not only unfamiliar with much of the observational material available, but uninterested in anything beyond a superficial glance at features that might be used to support his theory.

11.2.3   *Jeans and logic*

11.2.3.1   *Adjustables and independence*

Remember the Jeans Mass: the size of a condensation in a gas that will collapse under its own gravity. It's not a fixed number always and everywhere, but depends on things like the density and temperature of the gas one starts with.

Jeans calculates the Jeans Mass for a gas like air, at about room temperature, and a certain density; and finds it comes to within a factor of two or so of the masses determined for two galaxies (pp. 199-201). Proceeding along the same lines, using an overall density for the matter in galaxies determined by other astronomers along with a lower temperature, he finds masses about that of the sun (pp. 205-6). "And indeed there can be but little doubt that the process we have been considering is that of the birth of stars" (p. 206); that is, finding that his calculated masses match those first of galaxies and then of stars provides strong support of his overall theory.

In fact all it shows is that his formula can be adjusted to fit whatever comes along. He gives no reason for choosing room temperature as that of the protogalaxy, and indeed since "we can never know whether the hyopthetical primaeval nebula even existed" (p. 201) it appears he cannot give one. By choosing a different temperature one can make a Jeans Mass as big or as small as one likes. Since there is no restriction on the temperature one assumes, matching data proves — nothing at all.

Similarly, when he notices that the distribution of eccentricities in binary star orbits does not match what he expects as a final state (pp. 168-9), varying quite differently between close and wide binaries, he takes it to mean that his calculated effect of stellar interaction simply hasn't had enough time to work. By invoking three different mechanisms to produce binary star orbits in various ways (p. 225) he could explain almost *any* distribution of observations.

As a final example, he shows how well his liquid-star theory fits the Russell diagram (what we have called the Hertzsprung-Russell diagram), and finds "too much agreement remains to be explained away as mere coincidence" (p. 296); but again he has chosen a parameter in his theory to make things match (p. 298), and has had to ignore some known behavior of atoms (p. 297) as well as problems with the fit in certain parts of the plot.

The fact that Jeans can choose parameters to make his theory fit observation to some degree is not, of course, trivial. At the very least we can say his theory is consistent with facts, and if we can find some independent way to determine the parameters there is the possibility of actually testing the theory. But in these cases he has not done so.

The mention of independence brings up another habit of Jeans that must be brought out. In trying to sort out the ages of the stars, he finds it encouraging that all his methods of calculating seem to come up with a similar answer.

But we have already seen that his calculation of the ages of the stars, by equipartition of energy, depended on the assumption that stars lose significant mass through radiation and this affects their motions. But losing mass by radiation takes millions of millions of years, so a long time-scale has already been assumed. In seeking a way to power stars for this period, he rejects the fusion of hydrogen to helium (as suggested by Eddington) because it does not provide enough energy, and settles on his total annihilation idea. "On the whole, in whatever direction we try to escape from the hypothesis of the annihilation of matter, the alternative hypothesis we set up to explain the facts seems to lead back in time to the annihilation of matter" (p. 185).

So, far from having independant calculations agreeing, Jeans has several that depend on each other. They are *consistent*, perhaps, but actually come down to a single argument (if that) in support of the theory.

11.2.3.2　*Inconsistency and unfairness*

Further, there are bits of Jeans' work that do not fit together well and sometimes aren't even consistent. The local region near the Sun is more thickly populated with stars than the average (p. 85), and yet can be taken as representative (p. 256). One can argue that the regions referred to are different ones; but the first one being larger does not do away with the inconsistencey, and in *Astronomy as Cosmogony* (p. 9) Jeans notes that even very nearby our picture of stellar populations is incomplete.

On the method of energy generation in stars, "the annihilation of matter proceeds spontaneously, not being affected by the temperature of the star" (p. 291), and yet "[it is] highly probable that a star's rate of generation of energy depends on the physical condition of its atoms" (p. 311). The latter seems to be making a comparison between stars and not within a star, and to refer more to the ionization state of atoms than temperature as such, but the two statements need some thought before they can be made to fit. Jeans does suggest that a high enough temperature might affect radioactive decay (p. 139), which on Earth appears spontaneous ("whose rate as far as we know does not vary by a hair's breadth from one age to another" p. 150), and provides the model for his annihilation of matter reaction; in the bulges of galaxies he assumes a high temperature turns the reaction off completely.

In a different section of the picture, Jeans gives as support for his theory of the formation of the planets their fit to a cigar-shaped tidal arm pulled out of the Sun by a passing star (I've pointed out problems with the fit above). This hinges on the planets staying in the same order of distance from the Sun as they are now: Mercury nearest, etc. But he also requires that each form on a highly eccentric orbit, passing close enough to the Sun to raise a tidal tail in turn on the proto-planet to form its satellites. The eccentric orbits are then made circular through interaction with remaining diffuse material in the Solar System. It's hard to reconcile an orderly condensation of planets in the present sequence from the original tidal tail, one the one hand, with the violent changes in orbits required to pull out satellites, on the other.

None of these situations quite amounts to a contradiction, in the way that (for instance) Ball assumes in one place no dark stars at all, and in another an overwhelming number. It is possible to explain each of these situations in a consistent way. However, we can at least say that there is much in Jeans' picture that is not adequately worked out, and may yet make it untenable.

Perhaps more troubling, there is a sort of uneven treatment for Jeans' own theory and for competing ideas. The luminosity of a star depends on its mass, in that heavier stars are brighter (see Fig. 12.1). "Thus if we discard the hypothesis of the annihilation of matter, it becomes necessary to imagine some controlling mechanism, of a kind which would compel stars having the weight of the sun always to radiate at about the same rate as the sun ..." (p. 184). Although he adds, "There does not seem to be any general objection against supposing such a controlling mechanism to exist," he dismisses that proposed by Eddington and Russell for gaseous stars. My point is that, even with his annihiliation-reaction, Jeans *also* must find some reason for heavier stars to be more luminous. He probably did not notice, having already assumed a solution (his annihilation-reaction proceeds according to a sort of radioactive-decay law, and stars are born with a mix of quickly and slowly decaying atoms; p. 308), but the requirement is still there.

In comparing his tidal theory of the formation of planets with the competing one, that of independent condensation from the pre-Solar nebula, Jeans contends that the fact that the planets do not revolve in the same plane as the Sun's rotation makes the competing theory untenable; the theory "fails completely before this fact" (p. 240). The discrepancy is something like six degrees, not large though quite measurable. In asserting that his tidal theory of planets is therefore proven, Jeans makes an error we have seen before, the assumption that what has been thought of is all there is. The Laplacian theory of planet formation (the "nebular hypothesis") indeed required them to orbit in the same plane as the star rotates. Current theory has them forming out of the same cloud of dust and gas as the star, but out of an outer bit of it, which can have a somewhat different plane of rotation without much trouble.

Finally, there is a sort of "proof by assertion." In giving the physical background for his picture Jeans starts with the basis of atomic theory, that everything is made up of tiny lumps that cannot be subdivided further and still retain their identity. "It is easy for the scientist to say that, by subdividing water for long enough, we shall come to grains which cannot be subdivided further; the plain man would like to see it done" (p. 91). He then asserts that, given a sensitive enough weighing device, an evaporating glass of water would show jumps in its weight as individual molecules left. As an actual experiment this is quite impractical; no spring-balance or scale has remotely the sensitivity needed. But this is all Jeans gives the "plain man" in the way of demonstration: molecules exist because, if you could carry this out, this is what you would see — because molecules exist.

In summary, the *way* in which Jeans marshals support for his theory is troubling. He makes no distinction between rough agreement and exact matching, nor between independent proof and merely consistent calculations; takes infinitely adjustable numbers as vindication; is somewhat inconsistent in details of his theory, and occasionally unfair in the way he treats competing ideas. Taken together with his attitude toward observations, these establish significant room for doubt about his conclusions.

11.2.4 *Jeans and certainty*

How much doubt did Jeans himself have about his picture? Well, recall that at the outset he had characterized modern astronomy as confining itself to fact and not guesses. But fairly early on, when he discusses the number of stars in the Milky Way, he prefaces his remarks with "the whole discussion is rather in the melting-pot at the moment" (p. 65). In other words, he will not stay strictly to facts but will also set out some results that are not as well established. He covers something of the relativistic cosmology of Einstein and of De Sitter, starting with the cautious introduction that "Einstein claims to have established that space, although unlimited, is finite in extent" (p. 71). He points out that, "the general theory of relativity does not lead up to Einstein's cosmology in a unique way" (p. 75), that is, there are alternatives, de Sitter's among them. However, though the latter is seen as a possible picture, "there are many reasons against supposing it is a true one" (p. 80). He leaves the overall structure of the universe uncertain.

Jeans is also very realistic about the state of fundamental physical theory in 1929; the concepts are strange and difficult to picture, and "indeed it is rash to hazard a guess even as to the direction in which ultimate reality lies" (p. 119). Later on, "only a limited number of orbits are open to the electrons in an atom, all others being prohibited for reasons which we still do not fully understand" though more elaborate theories are coming to grips with the problem (p. 125). The latter are "groping after an understanding" but it may after all be a futile undertaking (p. 133). In this light Jeans' caution about "the present turmoil as to the fundamental laws of physics" (p. 157) seems well-founded. The Old Quantum Theory presented a great deal of promise, but also a great deal of confusion, and Jeans is convinced that its answers are incomplete at best.

What is Jeans sure about? as noted above, "there can be but little doubt" that his picture of the formation of galaxies and stars is the true one (p. 206). He has presented his picture of a primeval nebula giving rise to galaxies, stars, binary stars (by fission of liquid stars) and "sub-systems" (probably meaning planets); "... we need have no doubts as to its general accuracy, since observations confirms it repeatedly and at almost every step" (p. 226). So Jeans is quite uncertain about fundamental physics, but is sure of his theory overall.

Consider, though, one part of his picture, liquid stars. "Observational astronomy leaves no room for doubt that a great number of stars, perhaps even all stars," behave as rotating, gravitating liquid masses (p. 219); but he adds shortly afterward "we are totally unable to check our theoretical results by observation." I am not quoting Jeans out of context. Any straightforward reading of this paragraph would leave the reader hopelessly confused: observations simultaneously prove, but are quite unable to prove, Jeans' theory.

Jeans is not quite contradicting himself, but the way out is subtle and tells much about his ways of reasoning. Certain observational results, as interpreted by Jeans, provide (to his thinking) very strong support for his theory of liquid stars, to the point that he has no doubts about it. On the other hand, no one can actually see a liquid star in the process of fissioning to become a binary. This explanation, however, requires an insight into the theory and into Jeans' way of writing and thinking that are probably beyond the average layman reading the book. In that person, the suspicion should have arisen that Jeans' expressions of confidence and doubt are not to be relied upon. That, as we shall see, is true.

Rather late in the book, roughly a quarter of the way through the chapter "Stars" (that is, after the description of galaxy formation and evolution; star formation, and the tidal theory of planets) Jeans warns the readers, "Here we leave the firm ground of ascertained fact, to enter the shadowy morass of conjectrue, hypothesis and speculation. ... The reader who is hot for certainties may prefer to read something other than the remainder of the present chapter" (pp. 275-6). So, one might say, we have been well warned. But a bit later on, after Jeans has disposed of gaseous stars powered by temperature-dependent reactions, he says, "I do not think that much so far written in this book would be seriously challenged by competent critics" (p. 289). Once again there is a fairly stark disagreement between statements by Jeans: a "morass of conjecture, hypothesis and speculation" which would not, however, be "seriously challenged by competent critics." This time there is no subtle interpretation to explain the contradiction.

I think the only realistic conclusion is that Jeans is not terribly interested in rigor, in setting out what is firmly supported and what is less well established, and it shows up both in his theories and in his prose. He is *himself* certain that his picture is correct; parts of the structure underneath haven't been quite worked out yet, but that is a matter of detail.

The reader, however, is left completely confused as to what is certain and what is "conjecture, hypothesis and speculation." Under the circumstances, this is a good thing. If Jeans is not being careful about it, then there is doubt about anything he says; well-justified doubt, as we know now.

Let us go one step further. Suppose that we look at Jeans' book at a deeper level, beyond the juxtaposition of statements. Allow him to be a bit loose in this prose about exactly how far he his certain. He still seems to be definite about the state of fundamental physics. Whenever the subject comes up, he appreciates how little is known and how indefinite the results are so far. That much is uncertain, then. But his whole structure depends on a certain specific form for the matter-annihilation reaction. If it happens otherwise, his interdependent set of calculations must be set aside. This is the clue for the penetrating reader: the whole picture hangs on this one speculation.

11.2.5   *Clues for the layman*

I have spent a good bit of time on esoteric criticism of Sir James Jeans. The layman would not have noticed how his matter-annihilation power source was almost purely speculation, nor that his equipartition argument lacked a starting-point; and even the most skilled and trained scientists had trouble finding a problem with his calculation of the instability of gaseous stars. Along the same lines, a layman would not have pointed out that the spectrum of the bulge of the Andromeda Galaxy (which existed) would set his theory on solid ground or dispose of it.

Jeans does give (in a footnote) the equation for the Jeans Mass, so in principle the numerate reader could have noticed that by varying the assumed temperature one could make it come up with any quantity one wanted. This is rather like Herschel giving the formulae for Lagrange's calculation of the stability of the Solar System (recall Sect. 3.5.1.1): in practice it would take someone with the self-confidence of Bowditch to put in the numbers and actually show that the result is not quite as described.

Self-confidence would also be required to point out that the claimed match between the planets and the tidal-theory cigar really isn't very close, and to count up the single and multiple stars in the local region to show that Jeans' assertion of the rarity of the latter just isn't true.

A layman might have been confused, though, by Jeans' carelessness with his expressions of doubt and certainty, and thereby made a little skeptical. Based on that, a careful reading of the speculative nature of his matter-annihilation reaction and the realization of its importance to the whole picture could have raised enough doubt to match hindsight. Alternatively, a careful reading of Jeans' inconsistencies and unfairness might have led to the same skepticism. A reader whose skepticism had not been aroused, however, would retain a seriously flawed picture of the origin and evolution of galaxies, stars and planets, and thus of what Jeans considered the methods and results of modern astronomy.

11.3   The rehabilitation of Sir James Jeans

I have been rather severe with Sir James Jeans, but I do not think I have been unfair. The picture he presents in *The Universe Around Us* was never really accepted in detail, though parts survived, and the methods he used to support it and argue for it have serious flaws. He clearly started with his theory and chose his observations and physics to fit, occasionally doing violence to both. Indeed, of the astronomers I've looked at so far his techniques most resemble those of Sir Robert Ball, about whom I've struggled to find something good to say. Nonetheless, I intend to convince you that Jeans was neither a crank nor an incompetent, and indeed was a brilliant scientist whose contributions to astronomy (taken the right way) are impressive and lasting.

I do not need to argue the case with any astronomers. A simple perusal of *Astronomy and Cosmogony* shows the originality as well as the mathematical and scientific skill of Sir James. Indeed, the fact that my copy is a Dover reprint from 1961 (that is, thirty-two years after the original publication) and that much of its mathematical content is still taught in graduate astronomy courses attests to his contribution. But this is, of course, not of much use to the layman as a way of evaluating astronomy authors.

To explain how Jeans' work could be at the same time important and useful while still being seriously in error, I need to point out a few things about the book we are looking at as well as the state of astronomy and physics at the time it was written.

First, recall that *The Universe Around Us* is deliberately and expressly oriented toward the most recent activity of astronomy and physics. All the contents in several hundred pages of, say, Newcomb (1878) are squeezed into a couple of dozen pages toward the beginning. This means that solid results and well-attested theories are hardly mentioned, while ideas that are not yet firmly established are given much space. Recall the Second Irony: what is exciting may not be true. So instead of a book containing mostly reliable results, as a Newcomb or a Herschel might have produced, with a speculative chapter at the end, we have (in essence) the speculative chapter on its own.

Next, recall that Jeans was himself immersed in the new physics from the beginning. Instead of Herschel looking at spectroscopy and thermodynamics with caution and serious reservations, we find Jeans mastering the new techniques and extending them as far as he can push them. Imagination is a far more important and useful quality than strict rigor in exploring the new and the unknown. Thus, his embrace of the cutting edge skews the emphasis of his book even more.

Last, and possibly most important, astronomy was in a serious state of flux in the 1920s. In the Old Quantum Theory and Relativity it had been given tools that promised to solve old problems, as well as some questions that had hardly been able to be asked before. But there were no users' manuals issued along with these tools, and they themselves were of unproven reliability. There were enormous possibilities, but no clear directions. I will return to this situation in some detail in the final chapter; for the moment, I want to point out that desirable qualities in a scientist under these conditions — like creativity and the ability to do general, but non-rigorous, calculations — are not always those leading to sober evaluations of the reliability of final answers.

To these aspects I should add the utility of a competing theory. I have noted before the problems that can arise when there is only one suggestion to explain an observation. A competing theory, even if eventually proven to be incorrect, is invaluable in forcing a successful theory to be worked out in detail and in ramifications and with good rigor. It keeps the right answer honest.

So Jeans was extremely useful to the science, and to scientists. His effect on laymen is harder to evaluate; we will look at it again during the examination of another of his books, and at the end when setting out our final answers.

11.4   Summary and lessons

Sir James Jeans' *The Universe Around Us*, first edition, presents a picture of current astronomy and physics that is largely mistaken, though the problem sections are thoroughly mixed in with excellent insights and highly creative ideas. From the strict standpoint of hindsight, that is, of what eventually turned out to be the right answer, it would probably be best to discard it. I have argued, however, that there are other features of the book that make it well worth retaining.

There are sufficient clues in the treatment of theory, observations and logic to raise serious doubts about Jeans' conclusions in any reader who is familiar with the relevant science, or has the skill to check his results. However, the lay reader does not have the background to do this. A reader without such background, but with a certain initial skepticism and self-confidence, could have picked out problems with claimed agreement of theory and observation or at least with the logic employed in setting out Jeans' conclusions. Such problems, however, are subtle and must be looked for.

In this book we have met a number of familiar lessons for the scientist. Overinterpreted mathematical results appear, as well as inconsistency and unfairness in testing theories. I have mentioned Jeans' use of "we don't know of any" as equivalent to "there aren't any," which appears in several places.

New lessons learned for the scientist are:

- *Free adjustables are not support* for a theory. If your formula can give any mass at all by adjusting temperature or density, and you have no restriction on these, a match between your theory and observation is not a vindication.
- *Dependent calculations* provide no additional support for your theory. If you assume stars live millions of millions of years and calculate dynamics on that basis, a resulting time scale of millions of millions of years proves nothing new.
- By this time (1929) the *complexity* of astronomy had increased to the point that a flaw in a calculation could go undetected and unchallenged for a significant length of time. Even after a challenge was made, one could have the situation of two calculations reaching different conclusions with no clear way to choose between them. And astronomy has gotten no simpler over the years since.

For the layman we have:

- *When science is in flux* there will appear a lot of ideas that don't turn out to be true, or at least very accurate. It may be hard to tell when this is, but if books by different authors give very different pictures, or an author is very uncertain about fundamental science, this may be the case.
- *Claimed agreement may not be true.* If you notice that things are just not as they are described, you could be right. This takes a certain self-confidence.

Chapter 12

Sir Arthur Stanley Eddington, *Stars and Atoms*, Third Impression, 1928

The year is 1928. We have formally gone a small step backwards in time from the previous chapter. However, the scientific situation is for all our purposes exactly the same, and it is useful for some purposes to look at Eddington with some Jeans already under our belt.

By any measure Sir Arthur Stanley Eddington was one of the great astronomers of the first half of the twentieth century. Plumian Professor of Astronomy at the University of Cambridge, he was quick to understand Einstein's strange theory of General Relativity and to take charge of testing it. He also worked on the problem of the structure and evolution of stars, putting in place much of the current picture. Even today, in papers at the forefront of astronomy one will often find references to the "Eddington Luminosity", indicating a lasting contribution to the science. Much of his attention was given to the physics underlying astronomy, to the point that he could also be described as a theoretical physicist (though he himself prefers the title "astronomer", as we find on p. 83 of the book we look at in this chapter).

He also spent a significant amount of time and effort writing books for the public. We have at hand one of these, *Stars and Atoms* (Eddington, 1927). The first impression dates from 1927; we will use the third impression, carrying slight modifications, from the following year.

12.1 Motivation, audience and approach

This book originated as an Evening Discourse for the British Association at Oxford, that is, it began as an oral presentation for nonscientists much as did Airy's *Popular Astronomy*. Eddington took the opportunity to extend the original exposition for another series of talks and eventually for publication, so that something of the length limitation of a talk has been removed, but the flavor of an oral presentation remains throughout. The restriction on subject matter remains also; within the frame adopted, Eddington certainly cannot treat the whole science of astronomy. Indeed, "The treatment was meant to be discursive rather than systematic" (p. 5), so a complete structure is not to be expected.

Instead, he has chosen a theme, that of the connections between the atom (as recently revealed by the quantum theory) and stars. He thus concentrates on new results, as Jeans does, but in contrast with Airy (who was concerned with rendering believable something well-established). He will necessarily deal with theories and ideas that turned out to be incomplete, somewhat mistaken or perhaps completely wrong; his treatment of the uncertainties involved is a major concern of ours.

In contrast with Jeans, Eddington does not deal with origins. His only mention of where stars (in particular) come from is an assumption: "Granted that a star condenses out of nebulous material ..." (p. 107), a fairly general and plausible one. Speaking of the origins of various chemical elements he throws out an idea: "... unless indeed our universe is built from the debris of a former creation" (p. 106), but it occurs in a paragraph identified as speculation, in which he begins by saying, "We must await further developments" before anything more definite can be said (p. 105). Neither does he talk about galaxies, only noting that "Evidence is growing that the spiral nebulae are 'island universes' outside our own stellar system" (p. 9), and later saying something about the evidence (p. 93, in the context of the pulsations of Cepheid stars). He says nothing about General Relativity or Cosmology at all, which is something of a pity, for he worked a great deal in those areas, especially later in his career.[1]

Eddington is definitely speaking to a lay audience. He does use numbers without hesitation, both to impress and to allow comparisons (for instance, on the length of time various power-sources can supply stellar energy), but he requires no calculations of his reader and confines algebraic formulae to an appendix. His mathematical demands are thus about the same as Jeans, perhaps falling between those of Newcomb's two books. He admits that, "it is often necessary to demand from the reader a concentration of thought which, it is hoped, will be repaid by the fascination of the subject" (p. 5).

---

[1]Eddington did write popular, or semi-popular, books on these subjects. I have chosen not to look at them in this study mostly because they concentrate on physical theory and even the philosophy of science, too far removed from astronomy itself to be easily compared with the other books I've examined.

That is, the ideas are not familiar nor simple, and a layman will require some work to grasp them.

For those who are more capable, Eddington mentions that he has written up the same material with all the mathematics and physics left in, citing *The Internal Constitution of the Stars* (Eddington, 1926) published almost simultaneously. These two books thus form an explicit pair, much as I've taken Jeans' *The Universe Around Us* and *Astronomy and Cosmogony* to be. It is convenient to have both popular and technical expositions of the same material.

The word "fascination" above leads to perhaps the main motivation for the book: "if it has not lost too much in the telling, it should convey in full measure the delights — and the troubles — of scientific investigation in all its phases" (p. 6). As an explicit reason for writing, this is new to us. I suspect any astronomer who spends much time trying to communicate with the public is at least unconsciously trying to get across *why* he or she enjoys working in the science, but Eddington is the first we've seen to set it as a major goal. It is different from Herschel "trying to teach what we know" and Airy trying to explain how the Moon can be measured in Earthbound miles and pounds, and will affect what Eddington talks about and how he does it. At the same time, his account of the process of science echoes Newcomb's method in *Popular Astronomy*.

So Eddington's book has the short length and colloquial style of an oral presentation, with perhaps the danger of oversimplification due to constraints of time and listener concentration. He has the advantage of addressing one theme, like Airy, and so could in principle limit himself to secure and explicable parts of the science. However, he is working with the very latest results and trying to convey the process, which means he will have to present much that is uncertain or simply erroneous. From the point of view of hindsight he has set himself a demanding task.

12.2   Content and themes

Stars and Atoms is divided into three lectures, each consisting of subsections, as follows:

- **The Interior of a Star**: Temperature in the Interior, Ionization of Atoms, Radiation Pressure and Mass, The Interior of a Star, Opacity of Stellar Matter, The relation of Brightness to Mass, Dense Stars; 27 pages

- **Some Recent Investigations**: The Story of Algol, The Story of the Companion of Sirius, Unknown Atoms and Interpretation of Spectra, Spectral Series, The Cloud in Space, The Sun's Chromosphere, The Story of Betelgeuse; 34 pages
- **The Age of the Stars**: Pulsating Stars, the Cepheid as a 'Standard Candle,' The Contraction Hypothesis, Subatomic Energy, Evolution of the Stars, Radiation of Mass; 28 pages
- **Appendices**: Further Remarks on the Companion of Sirius, The Identification of Nebulium; 9 pages

As his title suggests and as he states in his preface, Eddington is concerned with a limited bit of astronomy: the structure and evolution of stars, as one can work it out with present-day (1927-8) knowledge of atoms (that is, using the Old Quantum Theory). His picture is one of stars gaseous throughout, even at extreme densities (with a couple of exceptions), a model he presents in some detail. Within the context of this model he interprets observations of many stars as well as some interstellar matter, then moves on to evolution. He deals with atomic theory and also quantum theory as it applies to spectra, so has some exposition of those areas. His gaseous stars are powered by the annihilation of matter, and almost all stars can be fit into a single track of evolution (so that the Sun, for instance, either has resembled or will resemble every other star in the sky).

There are three themes in the book that I would like to bring out. The first is a concern to convince the reader that one can indeed be confident about things that appear to be beyond all observation. "It is natural that you should feel rather sceptical about our claim that we know how hot it is in the very middle of a star ... Therefore I had better describe the method as far as I can" (p. 11). Similarly, "I wonder if there is an under-current of suspicion in your minds that there must be something of a fake about this photograph. Are these really the single atoms that are showing themselves — those infinitesimal units ...?" (pp. 17-8). This is the same as Airy's motivation, though of course the subject matter is different. Eddington is careful at many points to point out the links in the chain of reasoning so that it can be seen to be rigorous.

The second theme is also set out explicitly, that of the process of scientific investigation. The very idea of atoms was, not long ago, only a theoretical construct, and here are pictures of them (p. 18). Various ideas about the way stars might evolve are presented, two of them at least only to show that they were abandoned (pp. 106-9), and indeed, "It would be

difficult to say what is the accepted theory of stellar evolution to-day"
(p. 109). This is the same as Newcomb's program in *Popular Astronomy*,
though of course performed in a more limited fashion in a more limited
space. And though Eddington does not adopt Newcomb's explicitly historical approach, he does cover a significant period of time. In speaking of the
theory of gaseous stars he says

> This kind of investigation was started more than fifty years ago. It has been
> gradually developed and corrected, until now we believe that the results must
> be nearly right — that we really know how hot it is inside a star. (p. 14)

Here we have the theme of development and refinement set out, along
with caveats like "we believe" and "nearly right," whose use by Eddington
we will look at in a moment.

The third theme is the pedagogic one of the profuse use of similies and
methaphors. Of course these are useful in any exposition, especially oral,
and absolutely necessary if one is to describe mathematical science without
using mathematics. Eddington is a master of the craft. In describing the
high-energy interactions of atoms and electrons in stellar interiors he says,
"The stately drama of stellar evolution turns out to be more like the hair-
breadth escapades on the films. The music of the spheres has almost a
suggestion of — jazz" (p. 27). (Remember this is 1928. The reference
to films has aged well; but I should point out that jazz, in this context,
is a new, subversive, uncontrolled sort of music. The classical "music of
the spheres," by contrast, was slow, diginified, ethereal, more refined than
the ears of mortal man.) Similarly, a highly ionized atom turns out to be
much smaller than once thought: "Our mistake was that in estimating the
congestion in the stellar ball-room we had forgotten that crinolines are no
longer in fashion" (p. 39). For our purposes this theme could be important
if the metaphors get in the way of rigor, that is, if details of the verbal
picture are mistaken for the underlying physics.

There are two matters of Eddington's scientific philosophy that we can
now examine, having gone through (in particular) Jeans' book. The first
is his concern for a detailed, numerical fit between his theory and observations. To take one example: the luminosity of stars varies with their mass.
Eddington plots the observational data and runs his theoretical curve
through it, and it fits rather well. "Since I have not been able to give here
the details of the calculation, I should make it plain that the curve in Fig. 7
[my Fig. 12.1] is traced by pure theory or terrestrial experiment except for
the one constant determined by making it pass through Capella." (p. 35)

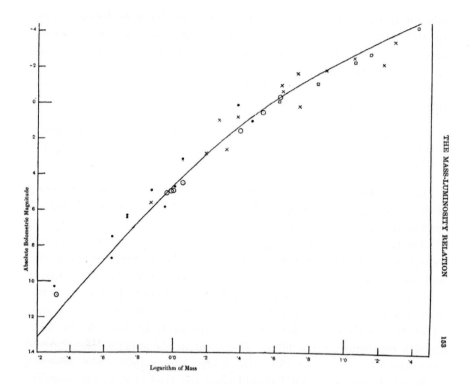

Fig. 12.1   The observed mass-luminosity relation for stars (crosses, boxes and circles) together with the theoretical curved derived by Eddington. The horizontal axis is the logarithm of the mass compared to the Sun, so that our own star appears at 0.0, and a star at 1.0 is ten times the Sun's mass. The vertical axis is in magnitudes, as we have seen before. This plot is taken from Eddington (1926); an identical version appears in Eddington (1927). Figure ©1926, 1988 Cambridge University Press, all rights reserved. Reprinted with the permission of Cambridge University Press.

So he makes plain his one adjustable parameter. Contrast this with Jeans, who in dealing with the same relation is content to assume a behaviour of his annihilation reaction that makes heavier stars brighter without going into any details.

I should make this a bit clearer. With his one adjustment, Eddington can move that curve up and down, and does so until it fits exactly one of his points. That it goes through, or close to, the rest is a victory for his theory and a strong indication that it contains much of what is actually occurring inside stars. It could have happened that the curve bent up too much at one end, or not enough, or was too close to a straight line, or too far from one, and all of this could have happened while still having heavier stars being brighter. Jeans does not begin to match his theory against observations to this level.

This is slightly unfair, in that Jeans' theory has not developed to the point where it can make numerical predictions. That leads to the second difference in philosophy between the two. Eddington proceeds by pushing *known* theory as far as it can go. In working out what happens inside a star, "there is no reason to anticipate anything which our laboratory experience does not warn us of" (p. 21), at least until forced to do so (p. 38). He acknowledges that it is possible that there is hidden in a star "something novel which will upset all your ideas" (p. 20), but "I am not abusing the unlimited opportunity for imagination" (p. 20). That is, it's possible to populate unseen or imperfectly-known areas of science with all sorts of creatures of one's imagination, but that is not the way to make progress. Contrast this with Jeans' technique of assuming things unless he knows of something to the contrary! Eddington here is much closer to Herschel, who would not allow an absorbing medium in the Milky Way because it seemed too arbitrary. Proceeding from a known theory allows Eddington to work out results in enough detail to match (for instance) the mass-luminosity curve, where Jeans simply doesn't have enough specifics to work from.

Along these lines it is characteristic, I think, that when presenting the unidentified spectral lines in some nebulae labelled "nebulium," Eddington is certain, "nebulium is not a new element. It is some quite familiar element which we cannot identify because it has lost several of its electrons" (p. 55). He is confident because all the elements up to number 84 have already been found; there is no place to put another one. He turned out to be right, as set out in the second and third impressions of the book: the nebulium lines were found to be due to oxygen and nitrogen (it is also characteristic of Eddington's prose style that he points out the strange and unidentifiable "nebulium" turned out to be — *air*). On the other hand, when Jeans was faced with the strange behavior required by his temperature-insensitive matter-annihilation reaction, he was quite willing to postulate unknown elelements (probably heavier ones, beyond Uranium at 92) and matter such as had never been detected on Earth (Sect. 11.2.1.3).

As a last example, Eddington concludes that stars must behave like ideal gases because the matter in them (highly ionized atoms) satisfies the conditions: much smaller than the spaces between them (this is the reason for the "crinoline" passage above) and with kinetic energies much larger than their energies of interaction. By contrast, Jeans bases his conclusion that they cannot be gaseous on his stability calculation.

12.3   Hindsight and Sir A. S. Eddington

12.3.1   *Caution and bafflement*

We have already seen an instance in which Eddington expressed his judgement about the reliability of his results. In the matter of calculating the central temperature of a star, the phrases "we believe" and "nearly right" appear. That is, he is not asserting categorically that there can be no doubt and no error, even though this is that part of the theory which he thinks is most reliable. To put words in his mouth, he would be surprised if the actual temperatures were much different; but there is always the possibility of the "something novel" of the previous paragraph. This sort of caution is characteristic of Eddington's book. Even after the numerical success of the gaseous-star theory he warns, "It would be an exaggeration to claim that this limited success is a proof that we have reached the truth about the stellar interior. It is not a proof, but it is an encouragement to work farther along the line of thought which we have been pursuing" (p. 40). Here his caution appears along with the theme of the development of astronomy, of the science as a process. To give another example, later on Eddington considers the relative abundance of elements in stars and thinks it "probable" that it is the same as on Earth; "All the evidence is consistent with this view;" but "this very provisional conclusion should not be pressed unduly" (p. 59).

With this kind of warning an alert reader would be reluctant to take anything as a final answer, and would not be surprised to find that here and there the material presented was in error. There are a few instances, however, in which Eddington uses less caution. Considering the two stars in the Algol system, "there can be no doubt that owing to the large tidal forces they keep the same faces turned toward each other" (p. 46). Hindsight finds no fault with this.

Another case, however, is more contentious. Dealing with the power source of stars, Eddington holds that "there can be little doubt that the release of subatomic energy is more rapid at higher temperature" (p. 116). Jeans was just as certain that the reaction had *no* variation with temperature. Both scientists, in this case, are talking about the postulated annihilation of mass, probably the combination of a proton with an electron (though Eddington is not as specific as Jeans); how could they both be certain about contradictory conclusions?

The subject of this disagreement is a process about whose details they had no direct evidence. Eddington may have started by extrapolating from chemical reactions and using the ideas of particle kinetics and thermodynamics (using the same sort of reasoning I used when presenting Jeans' gaseous-star stability calculation, as we saw in Sect. 11.2.1.1): that is, pushing known theory as far as it will go. More immediately, he is basing his statement on the fact that for a gaseous star to have an equilibrium at all, it must get hotter if you try to compress it.

Jeans, on the other hand, having shown to his satisfaction that a temperature-dependent reaction would make a gaseous star unstable, insists that annihilation must take no notice of temperature, and invokes radioactive decay as the model. Note that radioactive decay was quite inexplicable under the Old Quantum Theory. Here we have, I think, a clear clash of their scientific styles: Eddington pushes well-understood theory as far as he can, and a step into situations which are not well-understood; Jeans, overinterpreting a calculation, argues from analogy with something not understood at all. But both use confident words to express their choice; Jeans considers a reaction rate that depends on temperature "in every way contrary to the physical princples explained in Chapter II, as it is to all our expectations of atomic behavior." (Jeans (1928), p. 290)

This is almost the only case in which Eddington expresses what I think is an unjustified confidence in his conclusion, and it is arguable. His caveats are very protective against the action of hindsight.

Another protection is his willingness to be baffled. We have seen this before in his remark that there appears to be no generally accepted theory of stellar evolution. He refers to "quantum theory which is still the greatest puzzle of physical science" (p. 74), and on the apparently very similar central temperatures of very different stars, "The whole phenomenon is most perplexing" (p. 118). Indeed, he ends by saying, "I do not apologize for the lameness of the conclusion, for it is not a conclusion" (p. 121); that is, he is presenting a work in progress and does not pretend to know the answers yet.

Contrast this willingness with Jeans holding that an argument in favor of de Sitter's cosmology is its ability to explain the redshifts in the spectra of the nebulae, "while no other theory can explain them at all" (Jeans (1929), p. 79); and, in spite of the fact that his estimate of stellar ages should really be viewed with "caution, and perhaps even with suspicion," "Yet, if we reject it, so many facts of astronomy are left up in the air without any explanation, and so much of the fabric of astronomy is thrown into disorder . . . that we have little option but to accept it . . . " (Jeans (1929), p. 171).

Eddington is also aware throughout of the possibility of things he hasn't thought of. Considering the pulsations of the star Delta Cephei, "there are at least two alternative ways in which this heat could be supposed to operate a mechanism of pulsation" (p. 119), that is, there may be others using heat for a driving force, and (tacitly) maybe ways to drive a pulsation not involving heat. One can compare the technique we have seen before of saying, "there are two ways," showing one doesn't work, and therefore "having no doubt" that the other must be right (for instance, Jeans' exposition of the formation of the planets).

Eddington is aware of problems with the picture he has presented. Perhaps referring to Jeans' instability calculation, in showing one way in which Cepheids might be made to pulsate he notes, "this explanation is too successful . . . the trouble is that stars in general do not pulsate" (pp. 119–120). If stars evolve by radiating away mass, "Difficulties appear in the simultaneous presence of giant and dwarf stars in coeval clusters, notwithstanding their widely different rates of evolution" (p. 120). He acknowleges in each case the possibility that the problem may require modification or abandonment of his theory.

On a slightly different level, Eddington is not normally considered an observational astronomer (indeed, in some of his work it's hard to consider him an astronomer at all). But we have noted that he considers himself to be one, rather than a physicist. And he has a healthy skepticism about some observations; referring to the reported detection of the gravitational redshift in the companion of Sirius he cautions:

> I have said that the observation was exceedingly difficult. However experienced the observer, I do not think we ought to put implicit trust in a result which strains his skill to the utmost until it has been verified by others working independently. Therefore you should for the present make the usual reservations in accepting these conclusions (p. 53).

And indeed just previously he has given a situation in which an observation (the enormously high density of the companion of Sirius) was initially dismissed as "nonsense," until more and independent support for it could be found (pp. 50-1). It is instructive to compare Eddington's with Newcomb's attitude toward observations (Sect. 7.3.3).

12.3.2   *Problems*

So Eddington, by caveats and caution and a willingness to be baffled, has protected his readers very well against placing undue reliance on results that may be mistaken. Nevertheless, there are mistaken results, and it is worthwhile to take a look at them.

Eddington, like Jeans, thinks that stars are powered by the total conversion of mass to energy, a mass-annihilation reaction. He is not certain of this, however, saying in summary, "I do not hold this as a secure conclusion" (p. 121). But, since he holds to gaseous stars and thinks the reaction is sensitive to temperature, Jeans' reasons for annihilation don't apply; why, then, does Eddington think it probable?

It is required for there to be stellar evolution on the basis of mass. "...there can be no important evolution of faint stars from bright stars unless the stars lose a considerable part of their mass ...But if there is no annihilation of matter, the star when once it has reached the dwarf stage seems to be immovable; it has to stay at that point of the series corresponding to its constant mass" (p. 111). He proceeds to set out the choice in detail: either the stars live long enough to radiate away much of their mass, and so evolve into lighter stars; or their structure, luminosity and so forth are determined at birth by their mass. The postulated annihilation of matter reaction is the only way to power them for long enough for the first alternative to happen.

In hindsight, the question is why it seemed necessary to have stars evolve by mass in the first place. I think this shows something important about the way most astronomers think. Astronomical objects last a long time, so one cannot (except in very rare cases) catch one evolving. Astronomers therefore work on populations, assuming that if there is evolution, they can catch individuals at different stages. There is a great tendency to unify groups of objects this way. If one can unify *all* stars through one evolutionary sequence it would be a great advance; to stop partway, by saying stars are inherently diverse, seems somehow untidy. It's similar to the nineteenth-century tendency to see all nebulae as either resolvable or not, and Jeans' arrangement of external nebulae into an evolutionary sequence based on star formation at the edges. This assumption of similarity is sometimes true. In the case of stars, it is sort of half-true. The universe is untidy in places, and if one pushes a need to explain too far it can be misleading. In the case of Eddington, though, he does set out the alternatives and makes his motivation clear, so the reader need not be misled very far.

Hindsight also notices the complete absence of magnetic fields from Eddington's book, even when discussing prominences on the Sun, which he identifies as lifted by "violent outbursts" of the chromosphere (p. 71). It is now almost three decades since Newcomb mentioned such things in his *Astronomy for Everybody*, and it's worth some comment.

The two papers cited by Newcomb, Thomson (1901) and Cox (1902), were popular explanations of theories set out by the physical chemist Svante Arrhenius in 1900.[2] In the first, J. J. Thomson explains something of the behavior of "cathode rays," what we now know of as electrons. In particular, the Sun may be expected to emit them as a hot wire does, and they spiral along magnetic field lines. When they hit atoms they set up a visible glow. This explains the Aurora Borealis, and Arrhenius suggests something similar gives rise to the light of gaseous nebulae.

In the second paper, Cox explains (with some calculations) how radiation pressure becomes more important for smaller particles, and in particular how bits of comets can be pushed very strongly by sunlight, giving rise to their tails. He also shows how the Sun's corona could find an explanation through particles pushed by Solar radiation. However, these two papers are only an explanation of something unproven: "Such is Arrhenius' theory. It is too early, yet, to pronounce any judgement upon it ... [but] it is at least plausible" (Cox (1902), p. 278).

Indeed some parts I haven't mentioned don't fare well in hindsight at all, but the explanations of the Aurora and comet tails have survived. Eddington does not address the Aurora, but appears to have taken up the radiation-pressure theory for the corona and applied it to solar prominences. The puzzle is why neither he nor anyone else for decades put together the known magnetic field of the Sun with the known presence of charged particles there, and explained prominences that way. This would probably fit under the heading of "unexamined assumptions," or something that just hadn't been thought of.

On the other hand, Eddington thinks it "doubtful" that comet tails are the result of radiation pressure (p. 26), without any explanation. In the absence of any clue as to why he rejected Arrhenius' theory it is impossible to sort out a reason for this error. In sum, then, we can assign to Eddington some problems we have seen before, and a failure to forsee *everything*.

One observational problem was missed by essentially all astronomers of this period. The pulsating Cepheid stars had been identified as useful

---

[2]The paper by Arrhenius is in *Physicalishe Zeitschrift*, November 1900, but since I have not read it I won't formally cite it.

distance indicators, the period of pulsation being an indication of their absolute luminosity. Identifying them in distant clusters and other galaxies allowed the distances to these to be determined. The problem lay in the fact that there is another type of pulsating stars, RR Lyrae (or cluster) variables, that are significantly fainter than Cepheids. The latter were those first identified outside the Milky Way, giving distances that were too short by a factor of about two. Using this distance scale, for instance, the Andromeda galaxy appears to be rather smaller than the Milky Way. In fact, our own galaxy seems quite unusually large (as Jeans remarks in his book). But even here Eddington has a caveat:

> One cannot always be sure that what is true of the cluster stars will be true of stars in general; and our knowledge of the nearer stars, though lagging behind that of the stars in clusters, does not entirely agree with this association of colour and brightness. (p. 93n)

There is one lapse of logic that I want to point out. Eddington discusses the possibility of powering stars by the fusion of hydrogen to helium (though, as noted above, he thinks it less probable than the postulated annihilation mechanism). In recounting the evidence in favor of this reaction, he says, "To my mind the *existence* of helium is the best evidence we could desire of the possibility of the *formation* of helium" (p. 101, his italics). Here he is giving an argument for a process without a starting point, as Jeans did with his for the equipartition of energy among stars. If stars are born with nearly equal energies, no calculation of how long it might take to acquire them is valid; similarly, if stars are born with a certain amount of helium, its existence does not imply any formation of the element from other ingredients *in the stars*. Since Eddington (as we have noted) says almost nothing about the origin of stars, and nothing at all about the elemental composition of what they formed from, his argument is invalid (at least as a support for stars powered by hydrogen fusion).

12.4   Summary

Stars and Atoms is a highly reliable account of the state of stellar astronomy at the end of the 1920s, a field in flux and on uncertain foundations as far as atomic physics goes but one in the process of establishing what came to be the correct model of stellar structure and evolution. It is also a very readable account of some exciting and strange ideas that turned out to be true. Eddington is careful to sort out what is almost certain from what seems merely more probable than the alternative, and even manages to warn his reader against problems that were only worked out many years later. (The effects of the Cepheid/RR Lyrae confusion do not come into this study, as they were mostly confined to cosmology, but I assure you they were profound.)

His book contrasts greatly with that of Jeans, in the picture presented, in the confidence expressed in results, in the approach to scientific research. I have dwelt upon the difference at some length; certainly they were similar in being both highly capable and imaginative scientists; and if I had included one of Eddington's works on relativity and cosmology the difference might seem smaller. But I do not think I have overdone it.

Jeans was indeed more willing to imagine new processes and make things out of whole cloth, while at the same time seeking to leave no observation unexplained. Eddington pushed known theory into unknown territory, only reaching for something entirely new when there appeared to be no alternative. His work is rather less exhilarating, less full of unrestrained imagination; but far more likely to form part of the eventual reliable theory. It is ironic that a book written "to convey the delights — and troubles — of scientific investigation," and so ostensibly less concerned with the eventual truth of the results presented, turns out to fare better in hindsight than one more directly motivated to present the results.

In the process of science there is room for both approaches. From the point of view of the layman reading their works, the contrast can be confusing. Perhaps the best approach would have been to read both.

There are two lessons I can draw out of the comparison between Jeans and Eddington for the scientist:

- A *willingness to be baffled* when appropriate is a good way to avoid serious errors, especially when there is much uncertainty in the particular part of astronomy you're dealing with. This is to be contrasted with a *need to explain* too much. Of course, no scientist would be comfortable with the idea of *staying* baffled.
- Beware of the theory that attempts the *Grand Unification*, making all stars or galaxies (for example) essentially the same. Of course, all protons (for example) *are* the same, so one should not dismiss the idea out of hand, and some level of unification is what science is all about.

And for the layman, there is:

- If the author is not clear, or appears to be confused or contradictory in *expressing the level of uncertainty* of various results, they are probably pretty uncertain overall.
- If a "doubtless" result depends on a *doubtful result*, the least certain adjective applies.

Chapter 13

Sir James Jeans, *The Universe Around Us*, Fourth Edition, 1944

The year is 1944 and the world has changed again. It is still changing. Another war is underway, an even larger one, in which scientific and technical expertise is vitally important. The strange Old Quantum Theory has been superseded by the even stranger New Quantum Theory, hopelessly distant from anything imagined by nineteenth-century astronomers but supremely successful. General Relativity has its own brand of strangeness, though its meaning is still being worked out, and any applications to astronomical objects are either in the nature of small corrections to Newton or the rather speculative (even metaphysical) field of cosmology.

Sir James Jeans has now come out with another edition of his popular *The Universe Around Us*, the fourth (Jeans, 1944). Four separate editions in fifteen years attest to the popularity of the book, especially during the Depression and then war years when luxuries like accounts of astronomy are hard to budget, but perhaps more important is the pace of change in the science. From the beginning of this series Jeans himself has contrasted the "slow, arduous methods" of the nineteenth century to the "gold rush" of the twentieth (p. 37). An obvious bit of progress is the discovery of Pluto, between the first and second editions, but more and deeper advances have taken place by the time we get to the fourth.

13.1 The new edition

At first sight the fourth edition looks identical to the first. It has the same title and author, a similar length and an almost identical table of contents. The Prefaces to each of the four editions are printed, setting out the same general aim and audience, and the same philosophy as before.

There are seven chapters, with subsections as follows:

I. Exploring the Sky	84 pages
The Solar System	3
The Galactic System	5
Nebulae	3
The Distances of the Stars	7
Spectrum Analysis	6
The Photographic Epoch	1
Groups of Stars and Binary Systems	14
Variable Stars	2
Stellar Distances	5
Sounding Space	5
The Local System of Stars	2
The Rotation of the Galaxy	4
The Extra-Galactic Nebulae	9
The Structure of the Universe	1
The Theory of Relativity	12
Different Kinds of Space	6
A Model of the Universe	3

II. Exploring the Atom	53
Atomic Theory	2
Molecules	9
Atoms	10
Radio-Activity	3
Atomic Nuclei	2
Cosmic Radiation	4
Radiation	5
Quantum Theory	10
The Mechanical Effects of Radiation	5
Thermo-Nuclear Reactions	3

III. Exploring in Time	10
The Age of the Earth	5
The Sun's Radiation	5

IV. Stars	49
Stellar Observations	11
The Variety of Stars	19
The Physical Condition of the Stars	8
Stellar Structure	5
The Evolution of the Stars	6

If we note that the chapter on the Solar System consists almost entirely of identical material found in the "Carving out the Universe" chapter in the first edition, the books appear to have almost no differences. This is deceptive. There is a new subsection, longer than most and very important, on Spectral Analysis; the discussion of General Relativity and cosmological models has been greatly extended; more possible origins for binary stars are set out and discussed; and there is much more in the way of physical description of the planets.

Even more striking, though, is the complete removal of the annihilation of matter as a possible power source for stars, along with the whole of Jeans' liquid-star theory. The major difference between Eddington's picture and Jeans' has disappeared. Stemming from this, details of the possible origin of binary stars from the fission of liquid stars have been taken out, along with much of the description of the subsequent evolution of the putative liquid-fission binaries. In all, the additions and removals make for a slightly shorter book.

Roughly one-third of the fourth edition is either entirely new material, compared with the first, or is completely rewritten under the same headings. Rewriting has been done on large and small scales, sometimes replacing a word, a phrase, or a whole section, and much to Jeans' credit, the result is a seamless work. The problems I noted with the tenth edition of Herschel's *Outlines*, in which earlier sections were retained for no obvious reason or sit uneasily with later results, do not appear. (To be fair, Herschel's is a much longer and more ramified book; but the task of rewriting and updating an existing book so that it reads like a fresh start is still a difficult one.)

There are some subtle differences in Jeans' scientific style between the editions, which I will detail below. I have one stylistic observation to make before then. I have emphasized Eddington's use of metaphor and similie; well, they form a technique Jeans also uses often, though not quite as much as Eddington (the difference may lie in large part in the origin of Eddington's book as a series of lectures, a method allowing and requiring more informal ways of getting a point across). In comparing the different luminosities of stars, for instance, if the Sun is a candle, the faintest stars are fireflies, the brightest is a lighthouse, and supernovae are cities on fire (pp. 179–180). An atom in the million-degree center of a star attempting to retain all its orbital electrons is "trying to build a house of cards in a hurricane" (p. 193). Jeans is as apt and imaginative as Eddington in his pictures, if perhaps as little less profuse in using them (and he doesn't mention jazz).

In fact, Jeans has set out the basic reason for these rhetorical figures in the first edition. In order to describe quantum mechanics and, in general, things on the atomic level "It becomes necessary to speak mainly in terms of analogies, parables and models which can make no claim to represent ultimate reality" (Jeans (1929), p. 119). That is, for things "outside our everyday experience of nature" (Jeans (1929), p. 119), which I would add includes the very big as well as the very small, metaphor must be employed. For a popular audience the alternative of presenting the mathematics is simply not realistic. In the fourth edition he speaks of "pictures and models" instead of "parables and analogies," (p. 133, a sign that he has put serious effort into rewriting on small scales as well as large) but the passage is almost identical.

13.2   The picture changes

The fourth edition of *The Universe Around Us* sets out a signficiantly different picture from the first edition of the formation, evolution and structure
of stars. Part of the difference is the direct result of the continuing revolution in physics during the intervening fifteen years, and more is an indirect
result.

13.2.1   *Changes in background physics*

The main advance between the two Jeans books comes from the replacement
of the Old Quantum Theory by the New, along with some experimental
discoveries among the subatomic particles. Looking at the second field to
begin with, the atom had been known to be made up of the proton and
electron, of opposite charges but strangely different masses, and physicists
worked out how these two objects could be combined to make up the atoms
they knew. Then came the discovery of the neutron, a particle very slightly
heavier than the proton but having no electric charge (p. 118); it might
be considered to be an electron and proton bound together (p. 124). The
presence of a third type of particle complicates the picture; then came
along a fourth, the positron, a particle with the mass of an electron but
with a positive electric charge. It appeared that there were positrons in the
nuclei of atoms, but also electrons, making a simple picture of the nucleus
difficult to come up with at best. Then came the suggestion that electrons
and positrons might be created in pairs during the breakup of nuclei, and
thus might not actually be present beforehand at all; and that positrons
were not often seen because they would quickly meet up with an electron
and the pair would annihilate each other (pp. 126-7).

This sounds like the kind of matter-annihilation reaction that Jeans and
Eddington were talking about at the end of the 1920s, and so just what
Jeans might be searching for. But as the quantum theory developed it
became clear that a proton and electron *cannot* mutually annihilate; there
are rules about such things; only a particle and its antiparticle can do that,
and antiparticles are rare.[1]

Other reactions can happen, of course, and by 1944 some of the details
were becoming clear. The conversion of four atoms of hydrogen to one of

---

[1] It is still not certain why we live in a universe made up mostly of protons and electrons,
rather than one containing as many antiprotons and positrons. For almost all purposes
of basic physics there is nothing to choose from between matter and antimatter.

helium by the successive collisions of protons was outlined, as well as one
(requiring higher temperatures) proceeding by way of carbon, oxygen and
nitrogen nuclei (p. 149). In this way the almost speculative idea of getting
energy by fusion was given a detailed form, one that allowed it to be a very
definite candidate for the power source of stars.

So physics had ruled out one source of stellar energy and confirmed a
different one. The quantum theory had also advanced to the point that it
could predict in detail the spectra of atoms (and some molecules, though
those required much more calculation). Conversely, given a spectrum the
theory could provide much detail about its source. I must emphasize that
this was an enormous advance. Spectroscopy had been used by astronomers
for nearly a century, but for almost all of that time it depended on producing something in an Earthbound laboratory to match a celestial spectrum. Now, at least in principle (the actual calculations could be impractical for the resources of the time), astronomers could analyze something
quite beyond the capability of physicists to produce. This was a major
accomplishment.

As just one specific example of this, the spectra of very hot giant stars
(of class O and B) now yielded numbers for the physical and chemical conditions on their surfaces that completely ruled out Jeans' former suggestion
(Jeans (1929), p. 314) that they were nearby white dwarf stars. In general,
knowing the rules allowed less room for speculation and free assumptions.

Progress was also being made in the application and interpretation of
the two theories of Relativity. In particular, a few scientists were working
out how the entire universe must, or might, behave on the largest possible
scales. In his first edition Jeans had outlined the first attempts of Einstein
and de Sitter along these lines; the fourth edition has a much more extensive
section.

13.2.2 *Abandoning liquid stars*

The most striking difference between the two Jeans books is the complete
abandoning of the theory of liquid stars, powered by matter-annihilation
and lasting for hundreds of billions of years, occasionally splitting into binary stars by fission. What actually induced Jeans to change his mind so
completely is not clear from the book (I speculate that a Newcomb would
have been more forthcoming about the process); the new picture of gaseous
stars powered by fusion and lasting a few billion years at most is complete
and uncluttered. It might have been the demise of matter-annihilation as

a possible power source, or defects in any of several of the arguments Jeans had presented (and I have pointed out a few), or possibly the cumulative effect of many things pointing the other way. At any rate, liquid stars are gone, as are the age-calculations (such as equipartition of stellar kinetic energies) pointing to a very long time scale.

Jeans himself provides the corrections to some of his former problems. Here he is working out what, in a general sense, stars are made of:

> If we suppose that a star does not consist in large part of hydrogen, then the atomic weights of a number of stars, calculated on the supposition that the stars are wholly "gaseous," come out in practically every case higher than that of uranium, which is the weightiest atom known on earth. They not only prove to be higher, but enormously higher; so high indeed, as to seem utterly improbable.
>
> If on the other hand we suppose that the stellar matter may consist in large part of hydrogen, there is an alternative solution which requires that some stars at least should consist almost exclusively of hydrogen, while others contain a considerable proportion of hydrogen. There is nothing intrinsically improbable in this ... (p. 203)

Contrast this with his earlier contention that "we can disregard the possibility of a star consisting of hydrogen" (Jeans (1929), p. 280n), for which he gave no reason. (Eddington gives calculations for his stellar models in which he also excludes hydrogen, for example Eddington (1927) pp. 22-4. But he never says there are no hydrogen stars, only that the calculations come out differently for that element.)

In a later section of the first book, following on a paragraph phrased exacly as the first one cited just above with the exception of the mention of hydrogen, we find

> Again the explanation seems to be that the stars are not wholly gaseous. As soon as stellar interiors are supposed to be partially liquid, the calculated atomic weights are reduced enormously. They can no longer be determined exactly, but the atomic number of about 95 to which we were led from a consideration of the Russell diagram seems to be entirely consistent with all the known facts.
>
> Indeed other considerations seem to suggest the atomic numbers of stellar atoms must be higher than 92. *A priori* stellar radiation might either originate in types of mattter known to us on earth or else in other and unknown types. When once it is accepted that high temperature and density can do nothing to accelerate the generation of radiation by ordinary matter, it becomes clear that stellar radiation cannot originate in types of matter known to us on earth. Other types of matter must exist, and, as, with two exceptions, all atomic numbers up to 92 (uranium) are already occupied

by terrestrial elements, it seems probable that those other types must be elements of higher atomic weight than uranium. (Jeans (1929), pp. 304-5)

These passages, I think, illustrate very well a change in the reasoning used by Jeans between 1929 and 1944. Both logical chains start with the calculation that a gaseous star, to match certain observations of temperature and luminosity and having a certain response to radiation at high temperatures, must either be made mostly of hydrogen; or of something much, much heavier. The latter case requires elements well beyond any known on Earth. Initially, Jeans (having ruled out hydrogen for some reason) proceeds to reject gaseous stars, and still requires unknown elements to make his stars work. (Contrast this with Eddington, who proceeded on the assumption that there were no elements in stars that were not known on Earth: Eddington (1927), p. 55.) In 1944 Jeans has reexamined his assumption about hydrogen and no longer finds it useful (or, perhaps, convincing). That eliminates a complete chain of reasoning.

And now, as a byproduct, Jeans can say, "the whole universe appears to be built of the same 92 kinds of atoms as are known on earth," though he adds the caveat, "although it must be remembered that we have no direct evidence as to the kinds of atoms existing in stellar interiors" (p. 111). Well, Eddington was fond of pointing out that we have quite enough information about stellar interiors without having to see inside (Eddington (1927), pp. 11, 20). Perhaps the caveat is the lingering remnant of Jeans' former ideas.

Having accepted gaseous stars, Jeans is faced with his own stability calculation, in which he showed that gaseous stars with an energy generation mechanism that depends on temperature were unstable. He still thinks so: "the star would be highly explosive" (p. 207). He does not seem to have noticed, or at least accepted, Cowling's stability analysis (Cowling (1934, 1935)) that, to most astronomers, resolved the issue. Jeans gets around the problem by accepting a speculation of Eddington's that he dismissed earlier, (Jeans (1929), p. 299), that there is a time-lag between the rise in temperature and the increase in energy generation. Eddington himself was not at all sure he had found the answer, willing to consider stability an outstanding problem (Eddington (1927), pp. 120-1). Jeans sees confirmation in an analogy with laboratory experiments in radioactivity; in fact it is a false one. It seems he is still not cautious enough in accepting an explanation as long as it allows his theory to proceed.

Partly as a result of the modified theory of stars, Jeans is now willing to accept the possibility that not all stars are the same age (pp. 208-9), a requirement of his former picture that led to some rather forced assumptions. He is not sure of this, however, because it also leads to difficulties. These can be traced in part to a theory of stellar evolution that is flawed in detail, and to the fact that stellar evolution as we now know (especially in close binary stars) can be a very complicated business.

Jeans has also revised some of his accounts of observations and his interpretations of them. I have made a point in a preceding chapter that the data, even as presented by Jeans, are simply not consistent with binary stars being rare (as required by his former theory); he now notes that one count puts them at 68% of the more massive stars (p. 39).

I have pointed out Jeans' previous inconsistency in stating on the one hand that the Sun lies in a particularly dense part of the Milky Way, and then taking the local region to be a fair representation of the Galaxy. In the fourth edition he notes that the evidence for a local concentration of stars has become much weaker, so that it is not particularly probable (pp. 62-3); so the inconsistency has gone away, whether or not Jeans ever noticed it. Jeans did not say that there was no local concentration, just that the evidence for it had become "very uncertain;" this expression of doubt is one of several that distinguish the fourth edition from the first.

13.2.3   *Certainty and uncertainty*

In the first edition we found that Jeans was somewhat erratic in his expressions of certainty, to the point that a reader depending on them would have been bewildered about how much reliance to place upon many of the results given. There, for instance, after setting out his picture of the formation of galaxies by gravitational instability from a smooth beginning he says "It is of course at best only a conjecture that the great nebulae were formed in this manner — if for no other reason because we can never know whether the hypothetical primaeval nebula even existed — but it seem the most reasonable hypothesis we can frame to explain the fact that the present nebulae exist" ((Jeans, 1929), p. 201). But after proceeding to explain how he believed stars to form, he states "there can be but little doubt that the process we are considering is that of the birth of stars" ((Jeans, 1929), p. 206). His picture of star formation, about which he has little doubt, depends on his picture of galaxy formation, which he describes as only a conjecture.

In the fourth edition his description of galaxy formation as a conjecture is word-for-word the same, remaining a conjecture. But he has changed his conclusion about the picture of star formation to say, "And indeed it seems a resonable conjecture that the process we have just been considering is that of the birth of stars" (p. 223). He is not so sure now about his theory, or he is more willing to say so, or at least he is more consistent about it. I think this change in wording represents a significant, if subtle, shift in Jeans' thinking about the material he presents.

He is similarly less willing always to trust others' work completely; several astronomers "have suggested" that the star Beta Lyrae is of a particular kind, and "Struve thinks" that certain features in its spectrum can be explained by streams of gas (p. 50).

In the first edition he had rightly described atomic theory as highly uncertain, even in outline. In the fourth edition he refers several times to the successes of the New Quantum Theory: "For reasons which the quantum theory has at last succeeded in elucidating ..." (p. 142) and similar passages. But his descriptions still contain caveats: "So far as our experience goes" (p. 142), "little though we understand it" (p. 143). A note of caution has crept into his writing, along with a willingness to be baffled, rather than insisting on having an answer.

The latter feature, characteristic of Eddington, appears now in Jeans. In noting that the observed expansion of the universe is strikingly consistent with the cosmological theories of (for instance) Lemaître, he warns, "... it is difficult to hold our enthusiasm in check. Yet we must be careful not to interpret the concepts underlying the theories is too literal or too concrete a sense." (p. 88) Summarizing difficulties with reconciling his picture of stellar evolution with the apparently very different ages of binary stars, as well as the fact that all his suggested ways to form binary stars have strong objections to them, he concludes, "Clearly some piece of the puzzle is missing here." (p. 236) As to the whole picture of stellar structure and evolution, "We have to admit that there is still a good deal of uncertainly and even mystery about the whole problem, but we must remember that the whole subject is a growth of the last few years; order is emerging very rapidly out of chaos, and there is every reason to hope that all difficulties may be cleared up before long." (p. 209)

Contrast this, however, with Eddington's point that the behavior of gaseous stars had been under investigation for more than fifty years (Eddington (1927), p. 14; see Sect. 12.2). The seeming contradiction is a matter of emphasis, not actual disagreement, but I think points to a continuing difference in style between the two astronomers.

Jeans is content not to have the final answer on binary stars and looks to the future, not to confirm a theory about stars that he sets out, but to come up with the answer in the first place. This is a significant change in how he presents astronomy to the public.

13.3   Remaining problems

Not everything has changed between the first and the fourth edition, and there remain sections that present problems in hindsight.

Recall from the first edition that Jeans is sometimes a bit careless in his language. Further examples of this appear in the fourth: it is surely too harsh of Jeans, for instance, to characterize Newtonian physics as "meaningless and self-contradictory" (p. 78). His exposition of the theory of relativity and various investigations in cosmology that follow, however, are reasonably accurate. In other sections we do see problems with science.

13.3.1   *Carving-out problems*

Jeans still presents his picture of galaxies forming first, from a gas at roughly room temperature, though this time he guesses at a composition of a mixture of protons and free electrons (to make the masses come out right, p. 219); earlier he had not been even this definite. Stars are held to form at the edges of a slowly-shrinking galaxy, with the temperature adjusted to allow the condensation of objects of about stellar size (pp. 222-224). However, he is no longer sure that the central bulges of galaxies are made up of very high temperature gas, as called for in his earlier theory (which relied on some details of the assumed matter-annihilation process), and concedes that these regions may actually be made up of stars (p. 68). And he still stops the cooling, apparently arbitrarily, so that gas escaping from the edge of a shrinking star is not allowed to condense into planets. He is still certain about this result: "A mathematical calculation decides the matter definitely" (p. 242).

Having thus disposed of Laplace's particular theory, "and every other conjecture that attributes the genesis of planets to one star alone" (pp. 242-3), he concludes that two bodies have to have been involved. The sweeping reference to "every other conjecture" in itself is a warning sign, especially since we have found Jeans to be inexact in his statements on occasion. It comes under the familiar fallacy of "we know of none, so there aren't any." Note that this statement is *stronger* than in the first edition.

13.3.2   *The Solar System*

The carving-out problems bring us into the Solar System, where Jeans has changed almost nothing. Indeed, now he presents this theory of the origin of the planets as its own chapter. The very serious problems with his theory remain, particularly the difficulty in getting planets close enough to the Sun (within three radii!) to raise tidal tails to form satellites, and yet to stabilize them at their present vastly more distant places. He is still willing to take a rough consistency as exact agreement (pp. 248-9). The only significant change seems to be the concession that, when the planets were formed, the Sun was much larger and more tenuous (p. 252); which also means that planetary systems are probably more numerous than he previously thought.

As I said at the beginning of this chapter, Jeans has added to this edition a significant amount of observational and physical material about the planets. Though not an observer himself, he presents a commendable amount of skepticism about the canals of Mars (which he considers an optical illusion) and the interpretation of changes of surface features as vegetation responding to the seasons (pp. 269-70). Similarly, he points out that the finding of Pluto may in fact have been accidental, since the planet is not nearly as massive as the predictions required (p. 17).

Unfortunately, he is not at all consistent in this. Indeed, he describes reported features on Mercury as "permanent and clear-cut" (p. 263), when nothing observed from the Earth could really be called so. I suspect that he might have been led astray by the report of an enthusiastic observer. With this idea, he is more than ready to accept the conclusion that Mercury always presents the same face to the Sun, due to tidal locking (p. 257). Now, this was a widespread belief until much later (when radar observations disproved it), so Jeans cannot be held particularly responsible for the mistake. However, the result was based upon marginal observations that were difficult to make, so some expression of caution was certainly in order. It is hard to avoid the conclusion that Jeans was particularly willing to accept it because it fit in with his invoking of tidal synchronization in many places in the Solar System. Indeed, he states that as a result of this tidal effect from the Sun "Venus rotates so nearly at the same speed as the sun that it turns the same face to the sun day after day, and probably also week after week" (p. 257). How he can say that is a mystery, since he asserts (correctly) that the only thing we can see of Venus is the top of the atmosphere, never any features on the planet at all (p. 264).

Turning to a more distant planet, Jeans asserts that, based on certain photographs of Mars' polar caps, "the only possible inference would seem to be that the caps exist in the atmosphere, and not on the surface of the planet" (p. 268). I suspect the problem here is that Jeans is displaying his habit of dismissing a strawman, probably combined with an unfamiliarity with the details of photographic interpretation, but as he gives no further details it is impossible to be sure.

13.3.3   *Jeans and reasoning*

I've already noted some remaining or new problems with the consistency (at least) of Jeans' presentation and arguments. Along these lines, the "proof" by assumption of the existence of molecules also survives (p. 100). He is still happy with a rough agreement, asserting that the shapes of galaxies show that "observation conforms with theory" (p. 221) because they can be fit into his sequence, and he claims that red giant stars are "exactly" where they should be in the Russell diagram as young objects (p. 206). Apart from the difficulty of producing from the theory of the time anything that could be called "exact", the area occupied by red giant stars is large and does not have very sharp borders, so the use of that word is highly questionable. (This is aside from the fact that we now know red giants to be *old* stars.)

A minor, but I think indicative, instance of Jeans' continuing carelessness toward exact numbers is the assertion that the average distance from the Earth to the Moon is "rather less than 60 times the radius of the earth" (p. 29), compared with the figure of 60.27 radii used in a calculation a few pages later (p. 41). In the big scheme of things a difference of less than one percent may not appear to be much, but there is a contradiction here, and the later figure is pretty clearly chosen to make his calculation come out exactly right.

It is true that Jeans does not always claim exact matches between theory and observation. In a section found also in the first edition, he notes that the diameters of stars as found by interferometry are not a perfect match with theory (p. 171). And in the refined and updated section on the size of the Milky Way, he notes that "no great accuracy" can be claimed for some methods of estimation, and so the figure of 36,000 light-years from the Earth to the center of the Galaxy is in "satisfactory agreement" with a different determination of 50,000 (pp. 65-6). Compared to his general treatment, however, these cases are unusual.

(As a sidelight on an old matter, this section on the Milky Way contains new results showing its rotation (pp. 63-5), now firmly established in form although the numerical details are a bit fuzzy. Some of the evidence comes from proper motions of stars, some from Doppler analysis of stellar spectra giving line-of-sight velocities, and the answer is not what Mädler came up with a century before. Still, it is something of a vindication of the old German's basic idea, though his application of it was premature.)

There is an example of another problem with reasoning that I want to point out. It appears in the first edition, but is discussed a bit more in the fourth, and I think it shows an interesting way of obtaining the wrong answer.

In a section identical in both editions Jeans discusses the effect of the opacity of the material of a star.

> A star whose interior was entirely transparent could not retain any heat at all; its whole interior would be at a very low temperature and the star would be of enormous extent. On the other hand, in a very opaque star, all energy would remain accumulated on the spot where it was generated, so that the interior temperature would become very high and the star's diameter would be correspondingly small. (p. 201)

Jeans seems to be thinking of the following situation. We are given a certain flux of radiation coming from the center of the star. If the material of the star is transparent, that is, if it absorbs little or none of the radiation, a great deal of the material is required before the radiation is all absorbed. Conversely, if the material is opaque it takes very little material to do the same absorption. So if we require that a certain fraction of the radiation be absorbed, we need more transparent material than we need opaque material.

But this is not what happens in a star. There, the material is pulled down by gravity and held up by pressure, in part by radiation pressure. Opaque material will feel more radiation pressure, and so will be pushed out by it more. An opaque star is larger (all other things being equal). In fact, there are stars that pulsate because the opacity of a layer of material inside is changing.

By being a bit unclear about just what he is requiring, and also oversimplifying the situation (by leaving out gravity), Jeans has taken an initially correct line of reasoning and reached exactly the wrong conclusion.

In the first edition he uses this conclusion as part of his argument against gaseous stars (Jeans (1929), pp. 302-3). In the fourth it does not seem to lead to any further problems. Indeed, in his fully mathematical book (Jeans, 1928) he seems to have treated opacity correctly; it is implicit in his equations on pp. 118-9, though as he had no interest there in the detailed modelling of gaseous stars he made no comments about it.

13.4   Summary of the fourth edition

With the fourth edition of *The Universe Around Us*, Sir James Jeans has concealed under a superficially identical organization and with seamless writing an enormous change from the first edition. I have asserted the right of a reader to a well-written book, and not to have to deal with an obvious chop-and-change job. However, it is not easy to do a smooth rewrite; it takes effort and attention, and as we have seen even Sir John Herschel did not quite manage it. It is much to Jeans' credit that he has done so. It speaks well of him as a writer that someone picking up the fourth edition is only aware that there have been previous ones by reading the reprinted prefaces.

And that reader would only be aware of how much has changed by actually going through the various versions of the book. I do not know if I can get across just how difficult and unusual Jeans' accomplishment is in this area. He has taken an imaginative, one might even say brilliant, quite original theory; one on which he spent a great deal of scientific effort, worked out its features with elaborate mathematics, and defended against competitors over the space of decades; and — thrown it away. Such an action is against a strong feature of human nature, and unusual at this level of science; I can only think of one comparable case in astronomy. For this Jeans merits a great deal of respect.

Through the change in material Jeans' scientific style persists, although somewhat muted. While declaring that science has done away with guess-work (pp. 7–8), he presents a number of guesses. He expresses much more certainty than is warranted in many cases, though less than before. He still wants to explain everything, though now he is willing to be baffled at times. His presented results are subject to several problems that have been identified in previous chapters: taking a visual resemblance or a rough similarity as proof of a theory; overinterpreting calculations; dismissing a competing theory on the basis of a strawman or the "we know of none, so there aren't any" fallacy; and, in general, being more exciting than careful.

How much should a reader have trusted the fourth edition? Well, looked at from a distance with a bit of a squint, it's not too different from the current picture of galaxy formation and stellar theory. Galaxies do form from gravitational instability, a process now calculated using more elaborate forms of the theory originated by Jeans. The details are different: the sequence of nebula shapes as set out by Jeans is known to be false; the temperature and composition of the original material are much different; and stars do not form only on the edges of lenticular galaxies.

Stars are gaseous, largely made of hydrogen, powered by thermonuclear reactions (indeed, the specific reactions mentioned by Jeans), with lives of billions of years (instead of a hundred to a thousand times longer). The details as given in the fourth edition are erronous in places, especially dealing with stellar evolution (red giants are old, not young, stars). The reader is somewhat protected, however, by Jeans' caveats about the rapid growth and remaining puzzles in the subject (p. 209).

In all, if one takes closely to heart Jeans' caution that his presentation is only an approximation of the true picture (p. 7), and especially keeping in mind the notion of science as a process, the fourth edition is not too bad in hindsight. But Jeans' style still makes much of it unreliable, and a reader comparing the first edition with the fourth might be justified in seeing all the changes as indicative of a science that cannot make up its mind.

Chapter 14

In Summary: Reading the Astronomers, 1833–1944

It is now time to put together all we've found and try to come up with a real answer to the question with which we began. Greatly oversimplified, the short version comes in two parts: the good news and the bad news. The good news is that almost all that is written for the public by almost all astronomers is quite trustworthy, and the non-scientific readers need have few serious doubts about what they're told. The bad news is that wrong answers can appear even in the most conscientious authors without any warning or clue.

The rest of this chapter seeks to add detail to these short statements and provide some additional tools to the reader.

14.1 The character of science

Why would anyone, especially a scientist, write anything that is wrong, or that he's not sure of, anyway? I've already mentioned many particular reasons, but there is an overarching one, and it has to do with the basic character of science. As Newcomb emphasized in *Popular Astronomy* and any practicing scientist will tell you, science is a *process*, not a specific set of results. The mathematical methods developed for astrometry before and during the nineteenth century, then cutting-edge research, are now a computer subroutine invisible within modern telescopes themselves; they are no longer a subject studied by astronomers.[1] That is, as sort of a converse to the abandonment of exploded theories, this particular set of results is no longer part of research astronomy.

---

[1] Astrometry, that is, the precise location of objects on the sky, is very much a living field of research. My point is that the methods developed and used for a particular set of problems, like figuring out where to point a telescope to see a comet, have been worked out and assimilated.

Sir James Jeans uses the metaphor of the jigsaw puzzle to describe the process of astronomy:

> If [the information currently available about astronomy] must be compared to anything, let it be to the pieces of a jig-saw puzzle. Could we get hold of all the pieces, they would, we are confident, form a single complete consistent picture, but many of them are still missing. It is too much to hope that the incomplete series of pieces we have already found will disclose the whole picture, but we may at least collect them together, arrange them in some sort of methodical order, fit together pieces which are obviously contiguous ... (Jeans (1944), p. 13).

As a way to organize my list of lessons for scientists, I will push this metaphor farther than it was intended to go. I make a distinction between those lessons stemming from exploratory science, in which few or none of the pieces have been put togther; and those from dialectical science, in which there are competing ideas that have been developed to a significant extent and are being compared, that is when much of the sides and corners have been constructed as well as large areas of pieces firmly locked together.[2]

14.1.1 *Exploratory science*

The process necessarily starts with the unknown. At this point science is very creative: one must somehow generate ideas. I know of several attempts to make this a logical and predictable process, none of which work. One is looking for patterns in the data, unifying ideas, ways to make sense of a subject that may appear very chaotic or perverse indeed. Humans are excellent at finding patterns, even when there is really no pattern there. If the field is itself a new one, such as the application of Quantum Mechanics to astronomy in the 1920s, there may be very few indications at all of what might be a good line of investigation and what might be entirely misguided. When a field, or subfield, or area of scientific work is in this state,

- **Creativity is invaluable.** The ability to come up with a new idea, looking at the situation from an unprecedented angle or as a new construction, is the key. Whether it turns out to be right (in some way) or not is far less important.

---

[2]A distinction between two kinds or phases of science is not original with me. It forms a large part of the subject of the philosophy of science, and in its present form seems to have begun with Kuhn (1970). My distinctions are different, however, and my motivation is different, so I think it best not to try to connect to other work.

- **Rigor is difficult or impossible**, and anyway (I argue) undesirable. If you don't know how a matter-annihilation reaction might work, there's no point in trying to rule it out, or set hard limits on its behavior. And if you impose rigor too soon you can eliminate the right answer. Jeans dismissed the solar-nebula theory, in part, because there is a problem with discarding enough angular momentum for the Sun to contract into something as small as its present size. The problem is real, but it can be dealt with; Jeans imposed rigor too soon.
- **Suggestive calculations**, not fully worked-up models, have to be used to find one's way around the new landscape. They may not be right, in the end, but they give direction to the enterprise when nothing else is available. Jeans' original derivation of the Jeans Mass equation is sometimes known as "Jeans' Swindle," because of certain untenable assumptions made in it. The result, however, turns out to be correct. His stability analysis of gaseous stars turned out to be flawed, but stimulated better work on the subject.
- The exponent of a theory must be at least a little **irrational** in its support, in that he (or she) must have the confidence to carry on when it seems to be rather improbable. Especially with the complexity of theories and observations in the twentieth century (and the twenty-first), there may be much adjusting and extension of the original idea before it fits everything. At the outset, which things are a minor annoyances and which are insuperable obstacles are not clear.

Using our puzzle metaphor, we have a pile of loose pieces on the table and must set about doing something with them. Eddington is more inclined to find the edge and corner pieces, join them up, and work inward from there; Jeans will pull all the blue pieces together and see what kind of picture he can make. Eddington's style is more secure, but may take longer and can be blocked entirely by a few missing pieces. Jeans is more likely to see what the overall picture actually looks like — but it's also more likely to be a wrong one, when (for instance) he mixes the blue pieces of the lake with those belonging to the sky.

Coming out of this phase of astronomy, we have problems identified in previous chapters that led to erroneous conclusions:

- The urge toward **grand unification**, in which patterns are extended too far, imposing a simplicity that does not really exist. Eddington wanted stars to radiate away significant amounts of mass, so that all stars could be considered as lying along one evolutionary sequence; Jeans placed all spherical, lenticular and disk galaxies along one track. Closely related to this is the dependency of astronomy, especially during the nineteenth century, on **visual impressions**, which can be very misleading. Opposed to these is the hindsight virtue of a **willingness to be baffled** when things don't really work out or allow an explanation with the tools at hand.

- The **oversimplified calculation** together with the **overinterpreted calculation** both can come from trying to work out some kind of directions in a landscape without landmarks. I've asserted that it's sometimes counterproductive to insist on rigor too soon, when too little in known in detail, so simple calculations have their place; the problem lies in relying on them beyond their usefulness, or attributing to them a solidity they do not have. Related to them is the seduction of the **elegant mathematical result**, when things just work out very neatly in the mathematics and the scientist misses problems with the applicability of his calculation.

- **Adjustable and inexact agreement** are only to be expected when a field is just being explored. It is possible to demand too close a match between theory and observation when ideas are still not very well defined, a matter of imposing rigor too soon. The error lies in taking rough or adjustable agreement for solid confirmation of a theory, rather than for what it is: a lack of complete disproof.

- Finally, I want to emphasize the insidious **reinforcing errors**, in which two mistakes combine to make a more prominent error. In this connection, it is too easy to pick out something that appears to be confirmation of an independent result, without looking very hard at its origin.

14.1.2 *Dialectic science*

Eventually the picture begins to clear, and ideas are worked out in enough detail to be accorded the status of theories. There may be two or several competing explanations for something, or perhaps only one that seems viable. In any case, now the astronomer is seeking to put the theory on firm ground, and a somewhat different set of mistakes appears:

- It's mostly now that **unexamined assumptions**, the bane of a careful analysis, rear their ugly heads. Generally these are easy to see by hindsight, but very difficult to detect at the time, especially for laymen reading a scientific account. Related to this is the fallacy of **don't know of any means there aren't any**, in which ignorance is taken to be support or proof. It is a bit easier to detect the latter while it's in progress, though the clumsiness of adding a phrase like, "of which we are aware" means it's not always easy to keep in mind. Another related fallacy is the demolishing of the **strawman**, a simplified (or in some way not general) version of a competing theory, and thereby claiming all such theories are untenable. All these errors stem from things that are not thought of or not mentioned, a sort of "out of sight, out of mind" effect. As such they are most difficult for a lay reader to detect.

- **If you don't know everything, you may not know anything** is a terribly oversimplified way to say it, but does stick in the mind. It is possible to fit many of the aspects of a phenomenon with your theory, leaving a bit out, and think you're almost done; when in fact your progress is illusory. The Kelvin-Helmholtz contraction process seemed to fit the Sun very well, and certainly there was no other idea around for many decades (see the previous item); but no one had managed to figure out how to make the contraction put out a steady enough amount of energy.

- When things are going well, one is tempted to issue **categorical statements**, such as, "we understand gravity perfectly." It's probably safe to say that any such statement, standing without caveats or explanations, will eventually be found to be wrong. Simplicity has its price.

14.2 Rules for the layman

The previous section has attempted to organize the various ways astronomers have come up with to make errors, as they appear in the books under review. It's now time to look at the clues for the layman, things that could have or should have led to doubts about the material presented.

- First, **self-confidence** is the necessary factor before any other clues or ideas can be sought. The reader must be aware of the *possibility* of the author being mistaken, *and* the possibility of the reader finding the mistake, before it makes sense to look for it. Once this is in place, **confused or incomplete** arguments can be identified, **claimed agreement** can be scrutinized, and **illogical, inconsistent or unfair** lines of thought identified. I am not saying that any such defects in the presentation are necessarily signs of errors in the science, but I do hold that the author owes the reader a book without them. And remember that **"obvious" is your call.**

- Next, **read the book**. Notice the **doubts and caveats inserted by the author**, and take them seriously. Indeed, you need to keep them in mind more than some authors have done; as we've seen, a doubt expressed in one place may not be repeated (or, possibly, remembered) elsewhere. Consider that, if conflicting **opinions of other scientists** are reported, it means that the final answer is probably not in yet. And keep track of the **dependence of uncertainty**: if a particular result is labeled as "doubtless" but depends on a "doubtful" process, it's doubtful itself.

- There are some **warning signs** to indicate authors who are not to be trusted very far. Pronouncements made **out of their field** are, in general, no more reliable than any *you* might make (though there are, for example, such things as astronomer-historians). A stronger warning should be taken when the astronomer claims to understand another field **better than the experts**, when (for example) he asserts that the geologists or meteorologists are wrong and he (the astronomer) is right. A short and confident **dismissing of problems** in his theory, without any clue as to how they might be overcome, is troubling and may indicate the theory has fatal flaws. All astronomers writing for the public tend to use **numbers to impress** readers with vast sizes, distances, weights, ages; it comes with the subject. The method needs to be used with care, though, as it can have the effect (intended or not) of only showing how much more mathematically powerful the author feels he is. On the other hand, using **numbers to show uncertainties** and how well the theory matches observations is a good sign, showing a careful and sober author.

- The **state of the science** needs to be taken into account. A field that is just opening up will have much more uncertain material in it than an established one, and astronomers will have much less idea of what is important and what lines of thought will prove useful. The reader may have some clue as to the state of development of the science from other sources, or might be dependent on the author for guidance. And keep in mind the Second Irony: if a book concentrates on the **new and exciting**, more of it is likely to be uncertain, and the uncertainties will be larger.

- Even a very basic **skill with mathematics** can be enormously useful in finding where an author has made statements that are too sweeping or oversimplified. A suspicious reader could check Lagrange's formulae apparently proving the stabiliy of the Solar System, finding that it allows for the destruction of any of the smaller planets (as Bowditch did); and notice that Jeans' equation for gravitational instability allows adjustment to almost any mass one desires.

14.3   Applying the rules

Having gone through a long list of ways astronomers can generate errors
and layman can detect them, I have to emphasize that none of these are
infallible guides to what's right and what's mistaken. I have, for instance,
taken Sir Robert Ball to task for his criticism of Mädler's "Central Sun"
theory, even though Mädler's result was in fact wrong, and I have criti-
cized Sir George Biddel Airy for his comments on ancient Greek astronomy
even though Airy's Newtonian picture was much better science. I have
considered Jeans' use of the mass of M31 as support for his calculations of
gravitational instability to be at best a doubtful maneuver, even though it
is now generally agreed that M31 formed by gravitational instability.

There is no sure way of telling which of the particular results of astronomy at a given time are actually mistaken. Of course not; if there were,
astronomers wouldn't bother with mistaken results in the first place. The
best that can be done is to indicate when something *might* be erroneous,
when some doubt as to a conclusion is called for. I've given some ways
in which astronomers have reached, supported and argued for things that
turned out to be untrue; they have used similar methods for good theories.
I can only plant some doubts, which may or may not eventually turn out
to be justified.

At the same time, there are important and long-standing errors that are
difficult or impossible to identify without hindsight. The "volcanic" nature
of lunar craters, for example, would not have been seriously questioned
through the period covered by our books (with the exception of a note in
Newcomb's *Popular Astronomy*, which appears to be an isolated incident).
I've pointed out that it rests on the shaky ground of a visual impression, but
the importance of that appears only in retrospect. The flaw in Lagrange's
"proof" of the stability of the Solar System also appears mostly in retrospect, and only to people with the skill at mathematics required to detect
it. The Kelvin-Helmhotz contraction theory as a power source for the Sun
went without serious question for decades, even though it had certain difficulties in detail, mostly because there was no apparent alternative. And
for decades the explanations of the forms of solar prominences depended
on some kind of transparent atmosphere or radiation pressure rather than
magnetic fields, even though the necessary physics was at hand.

So applying my rules and indications will sometimes cast doubt on good
theories, while errors can go undetected. The best one can do is raise
suspicions that increase the probability of detecting a wrong answer.

14.4　Nowadays

I started out by saying that the analysis of popular astronomy by hindsight cannot be applied to current literature, for reasons that are still true. But the whole premise of going through this exercise for 1833-1944 is that some, at least, of the lessons learned can be applied to the science now. I will leave the intelligent application of these guides and indications to you (as, indeed, I must). But there are a few differences in the way astronomy is currently done that will affect how my results should be applied.

First, the calculations and theories of astronomy are now far more **complex** than anything in (or behind) the books we've looked at. I mentioned that Jeans' calculation of the stability of a gaseous star took several pages of algebra, and was oversimplified at that. A current calculation of the internal motions of a star includes the detailed quantum-mechanical behavior of many types of atoms, the transfer of radiation in a turbulent plasma, rotation and possibly magnetic fields, all done on a computer. It's just not possible to check such a thing by going through a bit of algebra (even if "a bit" means many pages), or by something as simple and straightforward as plugging a few numbers into Lagrange's formula for the stability of the Solar System. The reliability of any results from modern calculations is much harder to determine and rests on many hidden things. (Indeed, if I were looking at books published after about 1980 I would have to include a large, detailed section on how computers calculate things.)

Second, there are far **more numerical observations** that any theory has to fit. This means that many bright ideas die quickly, and any viable idea is worked up far more extensively before it sees the light of day.

Third, there are simply many **more astronomers** doing research than in the past. In principle this means that there are more eyes looking at a new result, and so less of a chance of some mistake slipping by. But the science is now much more subdivided than it used to be, so there are proportionately fewer with the background and skill necessary to examine any particular result. In the nineteenth century, for instance, one could assume that any competent astronomer could compute an orbit for a comet from a set of observations (including, if necessary, doing the reductions and corrections all the way from the raw numbers coming out of the telescope). Nowadays, an astronomer working on the spectra of giant stars would not necessarily be able to check the work of an orbit calculation (testing, maybe, the stability of the Solar System). Between the increase in numbers and the increase in fragmentation, I'm not sure which wins.

Connected with the increase in number of astronomers is the formalizing of the process of getting funding for one's reseach. I'm fairly sure that, for that and other reasons, astronomy is now a much more explicitly competitive place than it was.

One other aspect of current astronomy needs to be addressed. In several places I have referred to the current picture (of stellar structure and evolution, of the size and composition of the Milky Way) as the correct one. But how do I know that we've gotten it right now? Hasn't one lesson of *Hindsight* been that one firmly-held theory can be replaced by another, sometimes unexpectedly?

To deal with this question in any detail would mean doing what I've been avoiding from the beginning, making an evaluation of current science. My reasons for avoiding it still hold: it would be a large task, and my technique of hindsight in itself is not applicable. But it is a fair question to ask, given how critical I have been of other astronomers, and I will at least point the way toward my answer.

It's no good citing the consensus of many skilled and talented researchers, or the length of time a theory has been in place. These are unreliable indicators, and certainly on their own they should be unconvincing. Looking deeper, though, I think there are good reasons to be confident about most of the current picture, at least those parts I've labelled as "correct" in contrast with earlier ideas.

As a preliminary point, whether or not we've got it right currently, it is possible to say definitely that the earlier theories I've labelled as "incorrect" are indeed wrong. The Sun is not powered by gravitational contraction, for instance, and there is no metal rain or snow there. Stars are not made of liquid, nor powered by the mutual annihilation of protons and electrons. Wherever the right answer lies, it's not among the discarded ideas.

My confidence in the theories of nowadays, those I've called "correct" and used as points of reference for hindsight, is based mostly on their excellent fit to observations. An example relevant to several of the books I've looked at is the current picture of stellar structure and evolution, where I can point to an impressive match in many independent ways between observations and theory. There are areas that aren't well understood, especially with very massive stars and with details of formation, but the basic theory matches a wide variety of precise data. If a current model of the Sun, for instance, fails to match the observed speed of sound at a given depth[3] to within a few percent while reproducing exactly the observed luminosity, elemental composition, mass and size, the modelers consider it inadequate and search for problems. I assert that this sort of mismatch is different in kind from not being able to shrink the contraction-powered Sun at the proper rate to keep from frying or freezing the Earth. In that respect the theory is rather like Newtonian gravity in the nineteenth century.

But the mention of Newtonian gravity should put us on our guard: haven't I said that it turned out to be wrong? For all its successful precision, the rather different concepts and workings of General Relativity are the right ones — or at least more right.

It is in this sense that any of our current theories may turn out to be wrong. Gravitational force was replaced by a certain kind of curvature of space that within the Solar System (and indeed almost everywhere) had an almost identical effect. What we mean at the most basic level by "gaseous" and "nuclear fusion" may change. But stars will still be gasous objects powered by nuclear fusion.

I have been a little more cautious in my statements about current theories of the formation of galaxies and of the Solar System. For all that there is a broad and apparently well-based consensus on the basics of each, neither seems to me quite so firmly established as that of stellar structure, and there remains a real possibility of Eddington's "something novel which will upset all your ideas."

14.5   The usefulness of wrong answers

Well, what does it matter if an astronomer, writing for the public, gets it wrong? This is a subversive question to ask, here in the final chapter, but it's not a frivolous one. The answer depends on another question, one you must answer for yourself: why do *you* read astronomy?

If your aim is to understand something of the professional life of an astronomer, one of those for whom Eddington wrote when he intended "to convey in full measure the delights — and the troubles — of scientific investigation in all its phases" (as I've quoted from *Stars and Atoms*, preface), then the eventual status of any particular result in astronomy is secondary

---

[3]Unfortunately, explaining just how we know what the speed of sound is *inside* the Sun, as well as detailing all the other ways stellar theory and observations match, would take us too far afield and into very current research.

to the process itself. Indeed, it may be more interesting to see how an error comes about and survives for a time than to see a truth emerge, and not knowing the outcome adds some excitement. Certainly to understand the process is a good thing, as Newcomb in particular thought. This kind of writing has indeed led to books concentrating on the human story, sometimes written by authors with little or no understanding of the science itself. (They no doubt have a place as history, though it's not my favorite kind of writing.)

And I suspect that a large proportion of the astronomy-reading public is simply enchanted by the strange objects that astronomers find, as well as the stranger ideas they have about them. It becomes an exercise in mind-stretching and mind-bending, a very useful one sometimes, as we are made aware of things quite different and of new ways of seeing things, extending the possible. In this case the eventual truth or otherwise of an idea is not important at all, simply the originality of it. Here is where Jeans' liquid stars can still find a place.

But I do think that it is important, sometimes very important, for astronomers to avoid errors in their presentations to the public, even if those errors are only clear in hindsight. My reasons have to do with the nature of astronomy, and so for a moment I have to stray into the subject of the philosophy of science.

I have emphasized that almost all of Herschel's *Treatise on Astronomy* from 1833 could be taken over into a modern textbook without change. This is in spite of the great advances that astronomy has made in the interim, and in spite of the fact that the underlying science of physics has changed its concepts and language enormously. This is partly due to Herschel's careful qualification of his statements about the match of Newtonian gravity to observations; partly to his insight, which showed him that (for instance) he did not know the true nature of gravity; but also due to the fact that most of what he said was true, and still is. Similarly, Airy (when not commenting on the Greeks) and Newcomb wrote observations and inferences that are still true. There have been revolutions, but none of them has managed to "overthrow all our hypotheses."

Even when current physics was simply not up to the task of explaining something, as in the matter of comet tails in 1869 or the Aurora in 1878, Herschel and Newcomb could not only set out the observations but add reliable inferences to be drawn from them. I'd like to emphasize this as an invaluable skill: not only the careful delineation of where our certain knowledge stops, but the outlining of what can be inferred even in the absence of a full theory. When we lack the physics we are not always reduced to saying, "I don't know."

This means that we can read the old books not with the reaction, "that's what they thought at the time; we know differently now," but, "Ah! That's how they figured that out." It's the difference between a book of myths and a book of science. I think it is an important difference.

That is why I have been rather severe on Ball and Jeans. The former's methods tend to bring the process of science into disrepute (though he had some pretty bad results also); the latter was entirely too given to speculative results without caution or caveat (though he had some remarkable mistakes in technique also), at least for a book to be presented to the public. Imagination is invaluable for a research scientist, but the object is to find out something that is true.

And most of what astronomers will tell you is true. I hope I have given you something useful to figure out how true it is.

Acknowledgments

I wish to thank Virginia Trimble for much encouragement and help with bringing this book to life. I have received encouragement also from my colleagues at the Physics Department of the U. S. Naval Academy, especially Professors Elise Albert, Debora Katz-Stone and Jeff Larsen. Professors Trevor Ponman and Somak Raychadhury kindly found a home for an astronomer-errant at the University of Birmingham. Dr. Inga Schmoldt translated several journal articles from German, a skilled technical service for which she deserved gold or jewels, and was content with chocolate. Mark Hurn, the Librarian at the Institute of Astronomy at Cambridge, was extemely helpful in tracking down several publications from the late eighteenth and early nineteenth centuries. Dr. Avi Naim proofread the manuscript with scrupulous care, convincing me my typing skills are not what they were and making many useful suggestions. Kenny Russell's questions have been a motivation and sometimes an illumination to me, however inadequate my answers may have been. Drs. Scott-Morgan Straker and Sarah Tolmie have been my examples of highly intelligent, learned, interested people without a scientific or mathematical background, for whom I've tried to produce something useful. Finally, it has been a pleasure to work with Kim Reading, who drew many of the figures.

Bibliography

Airy, Sir George Biddell (1848). *Popular Astronomy: A Series of Lectures Delivered at Ipswich*, tenth edition, Macmillan and Co. (1881).

Ball, Sir Robert S. (1893). *In the High Heavens*, Isbister and Company Limited (1893).

Cox, John (1902). Comets' Tails, the Corona and the Aurora Borealis, *Popular Science Monthly* **60**, pp. 265–278, Jan. 1902.

Cowling, T. G. (1934). *Monthly Notices of the Royal Astronomical Society* **94**, pp. 768–782.

Cowling, T. G. (1935). *Monthly Notices of the Royal Astronomical Society* **96**, pp. 42–60.

Danielson, Dennis (2009). The Bones of Copernicus, *American Scientist* **97**, pp. 50–7.

Eddington, Sir Arthur Stanley (1926). *The Internal Constitution of the Stars*, Cambridge University Press.

Eddington, Sir Arthur Stanley (1927). *Stars and Atoms*, third impression, Yale University Press and Oxford University Press (1928).

Harding, K. L. (1828). *Philosophical Magazine*, July 1828, pp. 62–3.

Helmholtz, Hermann von (1863). *On the Conservation of Force* (Introduction to a series of lectures delivered at Carlsruhe in the winter of 1862–3), in *Scientific Papers: Physics, Chemistry, Astronomy, Geology*, Charles W. Eliot (ed.); the Harvard Classics, Harper and Brothers (1906), P. F. Collier and Son (1910), pp. 181–220.

Herschel, Sir John F. W. (1833). *Treatise on Astronomy*, New Edition, Longman, Rees, Orme, Brown, Green & Longman, and John Taylor.

Herschel, Sir John F. W. (1869). *Outlines of Astronomy*, American edition, P. F. Collier & Son (1902).

Hoskin, Michael A. (ed.) (1997). *The Cambridge Illustrated History of Astronomy*, Cambridge University Press.

Jeans, Sir James (1928). *Astronomy and Cosmogony*, Cambridge University Press.

Jeans, Sir James (1929). *The Universe Around Us*, Cambridge University Press.

Jeans, Sir James (1944). *The Universe Around Us*, fourth edition, Cambridge University Press (1960).

Karttunen, H., Kröger, P, Oja, H., Poutanen, M., & K. J. Donner (1996). *Fundamental Astronomy*, third edition, Springer-Verlag.

Koestler, Arthur (1959). *The Sleepwalkers*, Penguin (1962).

Kuhn, T. S. (1970). *The Structure of Scientific Revolutions*, second edition, University of Chicago Press.

La Place, Pierre Simon, marquis de (1799). *Mécanique Céleste*, volume I, translated with commentary by Nathaniel Bowditch, Hilliard, Gray, Little and Wilkins (1829).

Laskar, J. & M. Gastineau (2009). *Nature* **459**, pp. 817–819.

Laughlin, Gregory (2009). *Nature* **459**, pp. 781–2.

Loomis, Elias (1855). *An Introduction to Popular Astronomy*, Harper.

Mädler, Joachim (1846). *Astronomische Nachrichten* **24**, pp. 213–230.

Maxwell, James Clerk (1857). On the Stability of Motion of Saturn's Rings, in *The Scientific Papers of James Clerk Maxwell*, Cambridge University Press (1890), pp. 288–376.

Newcomb, Simon (1878). *Popular Astronomy*, Harper & Brothers.

Newcomb, Simon (1902). *Astronomy for Everybody*, Garden City Publishing Co.

Osterbrock, D. E. (1989). *Astrophysics of Galactic Nebulae and Active Galactic Nuclei*, University Science Books.

Russell, H. N., Dugan, R. S., Stewart, J. Q. & Young, C. A. (1945). *Astronomy, a revision of Young's Manual of Astronomy*, Ginn and Company.

South, James (1828). *Philosophical Magazine*, July 1828, p. 63.

Struve, F. G. W. (1826). *Astronomische Nachrichten* **97**, pp. 7–14.

Struve, F. G. W. (1828). *Astronomische Nachrichten* **139**, pp. 389–392.

Struve, O. (1851). Sur les Dimensions des Anneaux de Saturne, *Recueil de Mémoires Présentés À l'Académie des Sciences par les Astronomes de Poulkowa*, Primier Volume, l'Académie Impériale des Sciences (1853), pp. 349–354.

Thomson, J. J. (1901). On Bodies Smaller than Atoms, *Popular Science Monthly* **59**, pp. 323–335, Aug. 1901.

Tomkin, Jocelyn (1998). Once and Future Celestial Kings, *Sky and Telescope*, April 1998, pp. 59–63.

Trimble, Virginia (2008). Time Scales for Achieving Astronomical Consensus, *International Journal of Modern Physics D* **17**, 6, pp. 831–856.

Whiting, Alan B. (2007). Sir James Jeans and the Stability of Gaseous Stars, *The Observatory* **127**, 1196, pp. 13–21.

Index